HANDBOOK

of Weed and Insect Control Chemicals for Forest Resource Managers

HANDBOOK
of Weed and Insect Control Chemicals for Forest Resource Managers

By

Michael Newton
Professor of Forest Ecology
School of Forestry
Oregon State University

and

Fred B. Knight
Professor of Forest Resources and Director
School of Forest Resources
University of Maine, Orono

Timber Press
Beaverton, Oregon

Library of Congress Cataloging in Publication Data

ISBN 0-917304-25-X (Paper)
0-917304-63-2 (Cloth)

Newton, Michael, 1932-
Handbook of weed and insect control chemicals for forest resource managers.

1. Herbicides. 2. Insecticides. 3. Forests and forestry—Weed control. 4. Forest insects—Control. I. Knight, Fred Barrows, 1925-
II. Title.
SB951.N486 634.9'69 81-1629
ISBN 0-917304-25-X AACR2

PRINTED IN THE UNITED STATES OF AMERICA

Timber Press
Beaverton, Oregon

The use of trade names is for the information and convenience of the reader and does not constitute an official endorsement by the authors or the publisher of any product to the exclusion of others that may be suitable.

We dedicate this book to
Jane and Jane
Our Women Power

ACKNOWLEDGEMENTS

This handbook was prepared with the participation of many hands. Joanne Jones contributed not only accuracy but a touch of humor in her illustrations. Regina Pelletier and Julie Cone provided expert typing service and were indispensible in meeting schedule dates. For technical review, we are deeply indebted to Drs. J. D. Walstad, of Oregon State University, Corvallis, Oregon; H. J. Gratkowski, U.S. Forest Service (Ret), Roseburg, Oregon; D. M. Smith, Yale University, New Haven, Connecticut; C. H. Fitzgerald, University of Georgia, Athens, Georgia; and D. C. Allen, SUNY, College of Forestry and Environmental Studies, Syracuse, New York, whose contributions added greatly to the content and accuracy of this handbook.

TABLE OF CONTENTS

1. INTRODUCTION TO CHEMICALS CONTROLLING FOREST WEEDS AND INSECTS IN FOREST RESOURCE MANAGEMENT 1

Pests, pesticides, people and the future of forests 2
Forest pest problems amenable to pesticide treatment 5

2. FOREST VEGETATION MANAGEMENT 9

Vegetation as a management problem 9
Ecology of forest weeds 11

Habits of weeds 11
Crop tolerance to weed problems 13

Approaches to forest vegetation management and impact perspective 17
Herbicides and forest vegetation management 19

Labeling of herbicide products 22
Properties of herbicides related to effect 23
Formulation 24

Water-soluble liquids 26
Water-soluble powders 27
Wettable powders, flowable concentrates, granular and pelletized products 27

Oil-soluble liquids and emulsifiable concentrates 28
Unformulated products 29
Gaseous products 29
Matching formulation to use 29

Biological properties and uses 30

Phenoxy herbicides 30
Triazine herbicides 33
Phosphonoalkyl compounds 35
Organic arsenicals 36
Picloram and dicamba 37
Triclopyr 38
Asulam and amitrole 39
Dalapon and pronamide 40
Dinoseb 41
Paraquat 41
Nursery residual herbicides 41
Soil fumigants 42

Comparing the effects of herbicides 42
Silvicultural use of herbicides 42

Site preparation 52
Release 56
Timber stand improvement 60

Application methods 60
Foliage sprays 60
Dormant sprays 63
Basal and stem spraying 64
Stem injection 65
Stump treatment 67
Other methods 68

Uses of herbicides for wildlife habitat control 68
Safety 70

Minimizing the need for forest weed control 73
The vegetation management plan 73

3. FOREST INSECT MANAGEMENT 77

The insect problem 78

Forest insects 78
Economic injury levels 79

Use of cultural methods 81

Application 83
Bark beetles 83
Defoliators 84
Plantation management 86

Problem areas 87

Products insects 87
Shade tree insects 89
Seed orchards 89
Sugar bush management 90
Christmas trees 90
Wind break management 91
Recreation areas 92
Land reclamation 92

Types of damage 93

Leaf-eating insects 94
Sucking insects 98
Insects of shoots and roots 100
Cone and seed insects 101
Bark beetles 104
Other phloem insects 108
Wood destroyers and wood products insects 110

Insecticides and their application 113

Insecticide materials 114
Organic insecticides; pyrethrum, chlorinated hydrocarbons, organophosphorus compounds, carbamates, fumigants 115
Other insect controls; insect growth regulators, bacillus thuringiensis, sex pheromones, 119
Application 122

4. APPLICATION TECHNOLOGY 127

Properties of a spray mix 127
Formation of a spray 130
Sprayers and their components 131
Drift-properties of a spray 134
Application equipment 138

Power sprayers 138
Aircraft 139
Handheld equipment 142

5. RISKS AND CARE IN HANDLING PESTICIDES 147

Biological properties of pesticides 147

Insecticides 148
Herbicides 153

Potential for hazard 157

Human exposure 158
Non target species 161

APPENDIX 169

Management of Poisoning cases
Arsenic, arsenicals 169
Carbamate insecticides 171
Chlorinated hydrocarbon insecticides 172
Dinitrophenol group 174
Hydrocarbons (petroleum distillates, diesel etc.) 175
Lacrymators (soil fumigants, etc.) 177
Organophosphates 178
Phenoxy herbicides 180
Pyrethrum, Allethrin, Pyrethroids 182
Rotenone 183
Sulfur compounds (lime sulfur) 185
Triazines 186
General or unknown 187

Glossary 190
Index 218

CHAPTER I

INTRODUCTION TO CHEMICALS CONTROLLING FOREST WEEDS AND INSECTS IN FOREST RESOURCE MANAGEMENT

The forest is a place where trees grow, along with many other plants and the organisms that depend on them. The diversity of habitats, combined with disturbances, virtually guarantees continuous visitation by unwelcome plants and insects. Some of the insects, and some of the herbs, shrubs and trees, are pests that have appeared as the result of some event that created space or habitat for them. When there are too many, it is difficult or impossible to encourage the growth of desirable trees, and control becomes necessary.

The bounty of forest vegetation is an invitation for insects. The diversity is extremely great, and the insects do many important and useful things in the forest. Sometimes, however, they consume too much of a species or stand that has some special value or importance. Ordinarily, a manager will try to minimize the chances of epidemics, but sometimes damaging epidemics will occur in spite of conscientious efforts, and direct control measures may be necessary.

This manual is written for the individual who must become familiar with the important aspects of weed and insect control chemicals as used in the practice of silviculture. It is written at a level of technical detail aiming for brevity and clarity but in sufficient scope to be useful for teaching of principles or preparation of a pest management plan.

The manual is specifically directed toward proper use of insecticides and herbicides. It also addresses alternative methods of pest control in a general way. The emphasis on chemical control of insects and weeds does not imply that chemicals are the preferred method in all situations. It is an acknowledgement that pesticide technology is complex, and that the decision to use chemicals carries the responsibility to use them correctly. The technology of non-chemical pest control is fully as complex as that of chemicals because it often entails introduction of beneficial biological organisms or the application of energy-intensive cultural methods. A comparison of chemical and non-chemical methods requires as comprehensive technical treatment for non-chemical technology as for chemicals.

Pests, Pesticides, People and the Future of Forests

The potential for forest production in the United States is an asset of major economic and strategic importance. Climate, soil, access and market potential all are favorable for the economic management of timberlands on much of the half-billion acres regarded as commercial forest land. If these lands were all under reasonable levels of timber management, the timber produced would be roughly twice the total amount used in 1978, and the quality would be greatly improved.

But America is a net importer of timber products. Why? Among the many reasons, there are several prominent biological causes for the shortfall: the first of these is the widespread development of weeds and weed trees on cutover lands, largely as the result of human activity, which prevents the natural and artificial regeneration of the species being harvested. The next most important may well be the growth impacts and mortality caused by insects. A third cause is the widespread incidence of stem and root disease in desirable tree species. A diseased tree becomes a weed for all practical purposes, and the solution of many disease problems is dependent on various weed control methods through control of hosts. Thus it is that herbicides and insecticides have become widely used for reducing all sorts of biological losses, and their appropriate use can greatly increase net productivity of the overall resource.

Nearly all the pesticides used in forests have been developed for use in agriculture. There are literally tens of thousands of pesticide products being sold, but very few of these are registered for forestry use. Almost every aspect of pesticide prescription in forests is more complicated than comparable applications in agriculture because the complexity of forest communities and pest problems is so much greater. Having so few tools, the forester must be very sophisticated with the instruments available. The professional is in the position of a commanding officer confronted by an enemy with superior equipment and infinite numbers of troops — the strategy must be exactly right, and meager equipment and resources must be efficiently deployed.

Very few foresters are actually involved in managing the forested one-fourth of this nation. Their decisions affect the livelihood and enjoyment of millions. The land they are managing is regarded by a primarily urban public as a last frontier, and a wilderness to be neither entered nor altered by management. A common misconception is that each tree cut reduces the area of forests by that much; similarly, there is a popular belief that any use of a pesticide is a violation of the public's trust in that it inevitably causes irreversible harm. Facing these problems of public view, the forest pest manager must be very well informed about the biology of the pest problem, and the technology of control measures. Perhaps above all, the specialist must be able to reduce the complex biology and chemistry to simple terms and discuss it so that the public can be a constructive partner in planning. Recent history has repeatedly pointed out that certain elements of the public will insist on participation. The professional must be able to lead and educate in this process.

The pest manager must be knowledgeable about both pests and pest controls. A thorough knowledge of why pest problems occur and what conditions are appropriate for control is essential, along with an understanding of which pesticide to use, if any, and its anticipated effects on the forest as a whole.

The technical and social demands on the forest pest manager are heavy, and increasing. The forester is being required to give a scientific explanation of principles on which little training is available in professional schools. Until recently, basic scientific principles were developed and used chiefly during the research and development phases of pest management practices, and for pesticide registrations. Sound field biology and suitable labels and application technology were the principal ingredients of pest control planning. But today's forester needs a substantial technical background for competence at hearings and for preparation of environmental statements. Clearly, the practitioner needs a guide that will provide a practical compilation of vocabulary and principles in a single volume.

Thus it is that this book has come about. Its purpose is to bridge the gap between purely scientific theory and the empirical practice of managing pests. It is our intention to provide a structure of thinking that will beget a useful and defensible approach to the prescription process in resource management, and to tie to it some of the basic technology useful in making prescriptions work as planned. It does not attempt to offer a solution to each pest problem. Rather, it is implicit that each pest problem is a matter of professional evaluation in which the posing of certain questions leads to an understanding useful in developing a prescription or choice of prescriptions.

An important part of prescribing any forest practice is the accurate anticipation of its consequences. Every action in pest control, including the decision to take no action, has some effects. When developing a set of recommendations, the pest manager must be able to describe the set of consequences of each alternative proposed. Success will ultimately be judged in terms of whether the total outcome was, in fact, as intended. We will therefore offer a guide to evaluation of results and impacts and provide some insight into the general effects of major groups of compounds and non-chemical methods for solving certain pest problems.

The authors feel that the effectiveness of professionals in developing sound pest management programs is important to the long term welfare of the forest resource. Recent developments in pest management strategy and pesticide technology have given forest resource managers hitherto unavailable ability to maintain and improve the resource. These new tools offer a tremendous opportunity to tackle pest problems that have until now been difficult or unmanageable. Professionals now have the option (and responsibility) for pressing forward with progressive management programs to maintain and restore productivity of the forest resources. The general public expects a continual supply of the amenities provided by forests even if it does not demand them now.

This manual is a guide for the thinking that goes into pest management planning. It is not intended as a cookbook. Good pest management planning requires a great deal of professional dedication and study as foresters and wildlife managers add their knowledge of local biological situations to the relatively simple structure outlined in this manual.

Forest Pest Problems Amenable to Pesticide Treatment

Virtually every species of tree, many shrubs, herbs, vertebrates, birds, insects and other groups of organisms appear somewhere as pests. Conversely, the most important pest species may occur as non-pests. An organism is a pest only if it interferes in a substantial way with forest management objectives.

For each pest species, there is a point beyond which additional numbers are undesirable, and it becomes a pest problem. Small numbers of the pest species may not do much harm, whereas large numbers may totally foreclose the practice of silviculture. The professional must be able to evaluate whether the pest population is sufficient to be threatening or causing significant damage before any consideration of control is warranted.

Many pest problems are the result of inappropriate management in the past. Weeds, in particular, are dominant on many millions of productive acres from which high quality timber was harvested. The weed types have often been present for so long in some areas of productive site that the forest inventory shows unproductive weed cover to be the principal forest type. The original forest type may have been forgotten, but the growing potential for high quality timber is still present. Much of the eastern hardwood forest is of this type, converted from high quality hardwoods to low-grade or cull stands by exploitive harvesting. Extensive acreages are also in shrubs or low-grade hardwoods that were once high quality softwoods. One of the most important lessons in silviculture, as well as in pest control, is that there is no need for these problems to become rampant if harvesting and reforestation are managed carefully. Problem prevention through appropriate silviculture is the most desirable approach for minimizing future plant pest problems. Rehabilitation practices are necessary to correct situations resulting from past mismanagement. Timely maintenance usually is all that is necessary if forests are in good shape to start with.

Insect epidemics can also be affected by silvicultural practices. Severe insect damage generally happens only where conditions are suitable; epidemics require certain habitat conditions over large areas, combined with suitable weather. Not much can be done directly about weather, but stand management can influence microclimate, and maintain habitat so that epidemics are less likely. This is purely a silvicultural remedy. Ideally, in forests under intensive management many of the traditional pest problems may be avoided. Some of our very serious and chronic problems, however, have been caused by the introduction of exotics. These introductions have greatly complicated the management problem.

Intensive management may reduce some of the traditional and dramatic problems with pests but forests are never likely to be pest-free. The astute forest pest manager will undoubtedly identify local or specialized pest problems, each of which may be very important, but may not appear on many contiguous acres.

The forest manager must be a generalist to classify and handle the scope of problems on intensively managed forests. This will require fundamental knowledge of the basic relations between host and pest, or between site and tendency for weeds to prevail.

The cause of any pest problem is basic to its long-term control. The cause of a weed problem is usually related to some sort of disturbance that creates conditions especially suitable for the regeneration of the weed. An insect problem is often the result of a favorable composition of host species, but the causes of the composition, and the climatic and biotic control factors require careful analysis. It is often easier to kill insects and weeds than to prevent future outbreaks, yet the latter approach is preferable in all respects.

Weed and insect pests obviously differ in ways that are important in the planning and long-range execution of pest management plans; each requires its own approach. Weeds are of primary concern in regenerating forests or in stands that have been high-graded repeatedly. Clean-up of stands with poor species composition (i.e. timber stand improvement), is essentially a permanent solution of the problem, because continued proper management will prevent resurgence of the weed trees. Appropriate weed control during the regeneration phase will insure dominance of the preferred species, and weeds will thus be subordinate to the crop species until the next harvest. Insects usually become problems after desirable species are already growing, when the stands represent a substantial investment of capital and time. As long as there is a suitable host, a potential insect problem may remain. Habitat management to prevent insect outbreaks is thus a continuous program.

Unlike weed problems, insect attacks physically damage established desirable trees. The trees damaged or killed represent an accumulated investment that is either destroyed or rendered of lower value. If one creates an analogy to diseases in humans, an insect epidemic in a mature forest is analogous to a heart attack in an elderly person, insect problems of regeneration are like teen-age car accidents, resulting in some permanent injuries and some deaths. Weed problems, by analogy, are like pandemics and malnutrition among children. These differences in types of damage dictate differences in economic evaluation. Each group of pests is important. Control of each on a widespread scale is important to the quantity and quality of future timber supplies. Although weeds undoubtedly have the greatest impact on total timber production, insects may contribute greater dollar damage to the currently valuable stands. The forester may therefore be able to evaluate and prescribe for each class of problem as it arises, and couple management plans with public education programs describing proposed actions and the reasons behind them.

Public support for control programs is not always consistent with the scope of a pest problem. Weeds do not cause visible injury to dominant, valuable timber, and the need for their control is not widely understood. Weeds do not show up as a major and visible threat to future timber supplies because their impact is through lost growing opportunity. Insect control needs may be understood, but public perceptions about insecticides may not be germane to those proposed for use. Thus, a major educational program may be needed during program planning.

The newcomer to pest management can be overcome by the plethora of jargon used in pest management, toxicology and chemical technology. A Glossary of terms is provided in the Appendix for this reason. The seasoned practitioner may find it useful to clarify some terms. Each definition is preceded by a notation as to whether it is given in the context of pest control, chemistry, silviculture and so forth. There are some topics in which the only discussion of technical detail occurs in the Glossary.

CHAPTER 2

FOREST VEGETATION MANAGEMENT

Vegetation as a Management Problem

Undesirable vegetation is widespread in forests of North America. By definition, when vegetation is in conflict with the land manager's objectives, it becomes a weed problem. In this regard, forest weeds may be herbs, shrubs, woody vines or trees of any species that interfere with the manager's goals for timber, wildlife habitat, recreation or other values.

Vegetation may be undesirable in forests for a number of reasons. The most common is that certain species of no economic value occupy space and resources and reduce the growth of desirable trees. Weeds of this type may include superfluous individuals of desirable species, as well as grasses and forbs that compete for moisture and cause mortality in new plantations. Nearly all economic weed control is presently done for timber production, but wildlife habitat management opportunities are increasing.

Forest weeds are usually the result of human activities. The most common reason for the onset of a weed problem is that logging or abandonment of a field has led to the existence of unoccupied sites and germination of randomly occurring weed seeds. If weed seedlings are more successful than desirable species in the same location, the weed soon becomes the visible and dominant feature of that spot. In the event of widespread occurrence of such spots and general occupation by weeds, a weed or weed-tree dominated community replaces the trees removed. The result is a brushfield or stand of substantially reduced value to the landowner.

The act of forest weed control is the complement of harvesting. In much the same way as removal of the crop species favors development of weeds, removal of the weeds in systematic fashion favors development of the desirable species. Selective tree removal (thinning) for the benefit of residual trees is the basis for much of silviculture. Forest weed control is simply a group of silvicultural practices for controlling certain species to benefit others.

Weeds, whether they occur in forests, farms or gardens, have these characteristics in common:

1) Weeds are usually abundant after disturbance.
2) Weeds are most important when the crop is young.
3) Weeds, like crops, respond vigorously to fertilization, thinning and other cultural inputs.
4) Potential economic impacts of weeds are more severe on good than on poor sites.
5) Good crop management usually is at least a partial defense against weeds.
6) Weed control is of little value unless one has a well distributed stand of desirable species.

The above generalizations apply to all weed problems, including those in your vegetable garden. The following are unique to forests:

1) Some forest weeds can become very large and live as long as a forest crop.
2) Large stems and root systems, and thick bark, make woody weeds difficult to control.
3) Control of woody forest weeds must extend through a period of years until the forest crop becomes established.
4) Forest vegetation is generally complex; several genera are likely to contribute to the weed problem, and numerous species of plants are usually present without being weeds.
5) Wildlife and recreation interests may regard the weed cover as desirable; questions of land use and philosophy may be thus involved.
6) Control may be required only once or twice in each cropping cycle of 30-100 years; once dominant, crops tend to remain dominant. But the investment must be carried many years.
7) Topography dictates limitations on use of weed control equipment and methods.
8) Control operations on forest vegetation are extremely visible.

9) Forest weeds may preclude production of more desirable species for decades or centuries if not treated.
10) Noncrop forest vegetation is likely to be a major cause of disease and vertebrate pest problems.
11) Nonchemical methods of forest weed control tend to disturb soils and are limited in duration of effective control. They may stimulate other weed problems.
12) Methods of harvesting the previous crop have a strong influence on weed problems.
13) During the first several years, the crop species are almost insignificant parts of the community in terms of biomass.
14) Effects of woody weeds are most pronounced 3-30 years after establishment; herbaceous weeds restrict early survival of crop species.

Ecology of Forest Weeds

Habits of Weeds

Any vigorous plant may be the enemy of the regenerating forest. The species most important as pests are those with rapid growth and prolific abundance immediately after a disturbance; they shade seedlings and provide cover for foraging animals that feed on seeds and seedlings. The less vigorous plants that occur during this period, however, are of minor importance as weeds. In the regeneration phase, only the competing plant species with the ability to *dominate seedlings* are regarded as serious pests.

Domination of forest regeneration can occur by shading or by depletion of moisture, or both. Woody species that grow rapidly from seeds or sprouts are likely to be primarily a shading problem, causing mortality and loss of growth for many years after establishment. Hardy plants, especially grasses and low shrubs, are serious competitors for moisture for one to three years in areas of deficient summer moisture. Grasses that deplete moisture early in the summer are among the most important causes of mortality in new regeneration.

Weeds also cause losses in forest regeneration by producing physical injury. Aerial portions of tall herbs, such as ferns, and falling leaves, branches and stems from woody vegetation, have a severe impact on small seedlings. The use of large planting stock will minimize this problem, but it remains serious for natural regeneration and conventional nursery stock.

Very few stands of trees are ideal in composition or density. Hardwood trees of substantial size but poor form in hardwood and softwood forests containing higher value timber are among the most widespread of serious weeds. Superfluous desirable stems, culls of desirable species and noncommercial or low-value species are all weeds. The amount of wood in them represents an amount that would otherwise have accumulated on the desirable trees. Growth of weeds represents volume of unrealized potential on desirable trees. Many weed trees are the residue of a weed problem that earlier caused heavy mortality in regeneration, and still persist to compete with the surviving crop species. Removal of this group of weeds can be beneficial at any time up to harvest, provided the number and quality of desirable trees are adequate to utilize the site vacated by control of weeds.

When weed trees are the same size as the crop, there are occasional compensations for their impacts on growth, such as nitrogen fixation, pruning, etc. Some woody species, e.g. alders, black locust and *Ceanothus* spp. are nitrogen fixers. They make important contributions to soil nitrogen where deficiencies are severe. There is some evidence that nitrogen does not accumulate rapidly where supplies are adequate. Certain widespread shrubs of the *Ceanothus* group do not nodulate quickly, and do not fix large amounts of nitrogen for the first decade. Alder, *Ceanothus* and black locust are all strong competitors to tree seedlings when grown in dense stands. Whether these species are to be regarded as weeds depends both on whether they are fixing nitrogen to a useful degree and how dominant they are. At present, *Ceanothus* and alders seldom appear to be a net silvicultural benefit; black locust may have both commercial and fertilization value on sites where nitrogen is deficient such as coal spoils. Only rarely, however, can such a "nurse crop" be managed effectively because of competitive tendencies. They should not be established coincidentally with conifers unless it is certain they can be controlled or utilized.

High densities of larger cull or low-grade hardwoods may not be beneficial in any silvicultural sense. They distribute seed of inferior genetic origin and contribute little to the form of regeneration developing under them. They cause serious variability in stand size, density and quality. Being the dominant vegetation, they tend to have permanent control of site resources.

There are occasions when species occurring in association with forest regeneration may not compete severely, yet are undesirable for non-competitive reasons. For example, certain herbs with fleshy roots attract pocket gophers that can destroy ponderosa or Scot's pine plantations. The main benefit of controlling these plants may be the reduction of animal damage.

The keys to the competitiveness of woody weeds in regeneration are in their reproductive habits, root distribution patterns and juvenile height growth. Fig. 2-1 illustrates the different growth rates of several conifers and hardwoods with which they may be associated on average sites. Early sprout growth of hardwoods such as maples and hickories is greater than seedling hardwood or conifer growth. Sprouting species can be expected to appear as strong competitors after logging or site preparation. Certain of the light-seeded successional pioneers, such as alders, birches and poplars also have rapid juvenile growth from seed. All the species with such rapid growth are capable of suppressing conifers, and will cause heavy mortality or growth loss in plantations if they are abundant and occur in stands. Sprout growth of low-grade hardwoods is similarly able to cause severe losses in high quality seedling hardwood stands. It is important to recognize that a commercial species that attains large size at maturity may readily succumb to a competitor of rapid juvenile growth. Some low shrub species, such as snowberry or manzanitas, cause severe soil moisture depletion in the root zones of regenerating conifers.

The growth dynamics of young stands are one of the principal components of forest succession. The emergence of successive groups of plants as dominants is determined by their success in becoming established, in becoming dominant over other plants, and in reproducing. Succession in forests subjected to selective removal of conifers and other species with limited sprouting ability, however, leads to an abnormal abundance of sprouters, as understory species and cull hardwoods become dominant. After being projected into dominance, these species have a strong tendency to remain dominant and to form highly stable sub-climax communities which are atypical of the site in the absence of logging.

Crop Tolerance to Weed Problems

Seedlings of forest tree species differ in their abilities to tolerate suppression by hardwoods. Species intolerant of shade must be substantially dominant to avoid major loss of diameter growth, and equal to competitors in height to avoid loss in height growth and mortality. Species of intermediate shade tolerance avoid severe losses in diameter growth and are little af-

Figure 2-1. Height-age curves for several open-grown conifers and their approximate equivalents for hardwoods.

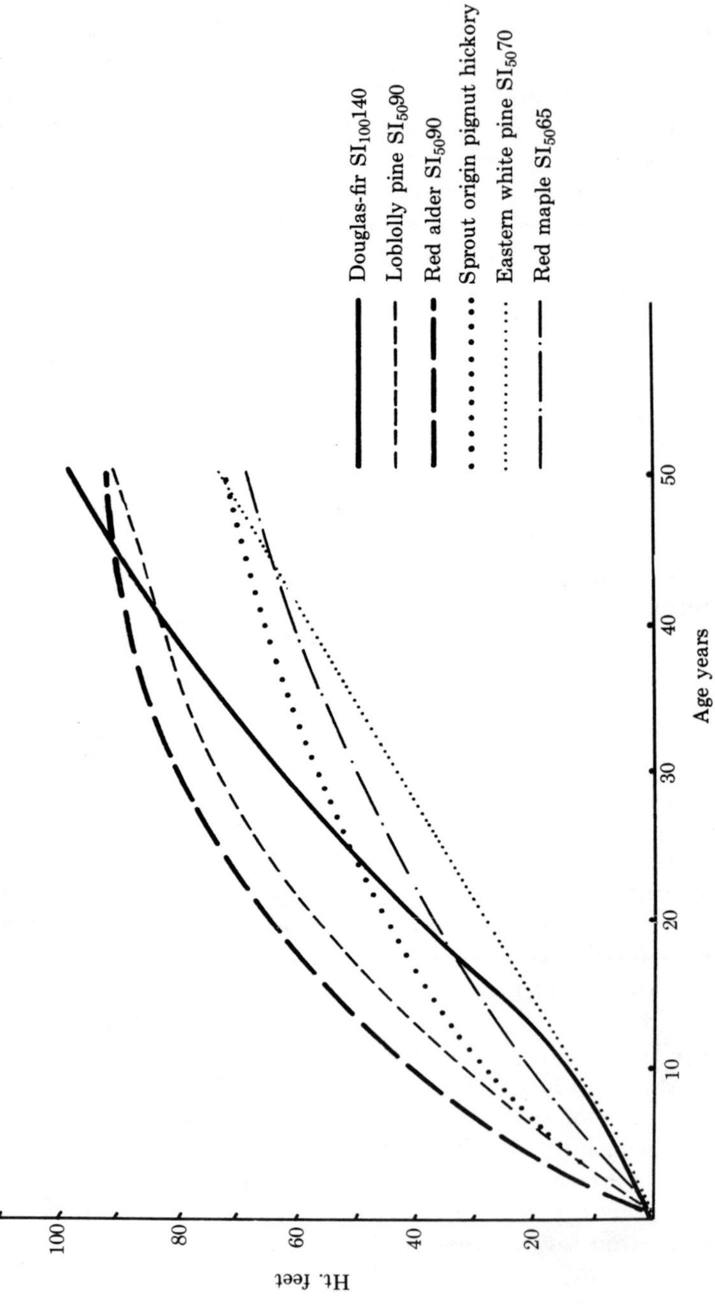

Table 2-1. Tolerance of commercial conifers

Tolerant or very tolerant	Intermediate	Intolerant or very intolerant
All hemlocks	Eastern white pine	Eastern redcedar
All spruces	Western white pine	Red pine
The white firs	Sugar pine	Pitch pine
(balsam, grand,		
white, subalpine,	Atlantic white-cedar	Shortleaf pine
silver)		
Thuja spp.*	Slash pine	Loblolly pine
Incense-cedar	Monterey pine	Pond pine
Alaska cedar	California red fir	Sand pine
Coast redwood*	Giant sequoia	Jack pine
	Douglas-fir**	Virginia pine
		Longleaf pine
		Ponderosa pine*
		Jeffrey pine
		Lodgepole pine
		Larches
		Noble fir
		Junipers
		Douglas-fir (Coastal)*

* Preferred forage for deer and small mammals.
** Except coastal strains, which are intolerant.
Condensed from: Baker, F. S., 1949. A revised tolerance table, J. For. 47:179-181.
Fowells, H. A., 1965. Silvics of forest trees of the United States, Agr. Hdbk. 271, Forest Service, USDA, 762 pp.

fected in height if they are co-dominant with competing species. Tolerant conifers may continue to grow without loss of height increment when somewhat below the heights of nearby trees, but severe losses in both height and diameter growth occur if they are below the live crowns of the hardwoods. Thus conifer species grow most rapidly in the absence of significant competition. The *degree* of loss in growth is somewhat related to tolerance as well as density of cover. Mortality is strongly related to tolerance. Table 2-1 lists the common conifers according to their tolerance to shade. When reforesting, selection of the more tolerant of desirable species will minimize losses due to competition with the least amount of weed control.

Tolerance to shade does not imply that a coniferous seedling is in a favorable habitat in the shade – tolerance is a term that describes probability of mortality in an understory position. All commercially important conifers and hardwoods grow best when dominant. They are also less subject to certain other environmental hazards when competition is minimal. One of the major determinants of tolerance is the ability to withstand the impacts of falling litter from overhead vegetation. The shading effects of the overstory cause slender, fragile stems and reduced growth; the falling litter bends, breaks and otherwise damages suppressed regeneration. Some species, particularly conifers with *flexible tops,* tolerate such damage better than others.

Another important feature of ability to tolerate competition is animal damage. A dominant cover of hardwoods and shrubs 3-15 years old is typically the very best cover for deer and several important species of small mammals, (rabbits, hares, mountain beaver, etc.). Suppressed trees are also likely to be subjected to severe damage by animal feeding. Suppression reduces the ability of a seedling to recover from such physical damage. Browsing and clipping remove current growth, especially, and contribute seriously to the suppression problem. Removal of the cover may not eliminate animal damage entirely, but it will improve vigor, hence ability of the trees to tolerate a given degree of damage, and to grow out of the range of susceptibility sooner.

Forest weed problems may be very long-lived. Natural regeneration does not readily invade old pastures, meadows and other areas fully occupied by herbs, in areas with dry summers, or shrubs in areas with moist summers. Plant succession is very slow on such areas, because of the extremely droughty condition set up by herbs accustomed to completing their growth cycles early in summer. A conifer invading a dry site will have to endure severe soil moisture stress through the hottest part of the growing season without significant available moisture. First-season mortality takes such a toll that forest cover does not become visible until after hardy shrubs or hardwoods have paved the way; those hardy species also suppress the conifers for many years before the conifers eventually emerge as dominants. When trees do appear, they are usually scattered, inadequate in number and have stems of poor form. In moist climates, shrubs limit availability of light.

Patterns of forest succession vary among forest types; desirable species do not always succeed pioneer vegetation naturally. Red alder in western forests, for example, invades cutover Douglas-fir lands. Instead of eventually giving way to Douglas-fir, a more common pattern is to give way to salmonberry, which may last several hundred years as a shrub type. In these forests, as in those of eastern white and southern yellow pines, an abundance of seed at the time of disturbance (e.g. harvesting), often insures an adequate population of conifer seedlings. Determination of whether the site will be dominated by these conifers depends largely on presence of pioneer or residual shrubs and hardwoods. If brush is prevented from suppressing the young conifers, the softwood type prevails. This principle is extremely important in the weeding of coniferous forests; it is used in reverse on powerline right-of-way, where shrubs may be encouraged as a means of *excluding* trees.

As a general rule, forest seedlings need unoccupied spots in which to survive and grow vigorously. Preparation of such weed-free ground for reforestation is known as *site preparation.* For best growth, trees often need early selective weeding, called *release.* Without either site preparation or release, the establishment of a forest crop is often delayed for decades, and sometimes centuries.

Approaches to Forest Vegetation Management and Impact Perspective

The perspective in which a weed problem is viewed is largely a matter of landowner objectives. The approach taken to deal with forest weeds depends on the kind of vegetation involved, availability of treatments known to be effective, degree of financial commitment, topographic limitations and legal or regulatory restrictions.

Some weed trees have potential for utilization. Under such circumstances, problem vegetation can be harvested at a profit. Landowners with pulpwood-using facilities may find it economical to continue to grow the weeds for later harvest. As a general principle, the negative value of a weed problem is determined by its impact on the growth of the highest priced product that can reasonably be expected from the site. In some regions, trees that won't reach sawtimber size and form have little value to the landowner who manages for maximum total return from the woods. For these owners, any tree that doesn't have the potential for producing premium quality forest products is a weed if it competes with the higher quality material.

Whether a given plant is a weed depends on the situation in which it occurs. For example, a row of stately sugar maples along the border of a spruce-fir woodlot in New England has considerable aesthetic value due to brilliance of fall colors. Maintenance of such amenities has substantial public relations value. In the spring, they may even be offered for sugar production for full resource utilization and for goodwill among neighbors. Similar consideration may be given for game habitat spots, stream borders and scenic outlooks, provided they do not interfere substantially with over-all management objectives. Where interference is substantial, there is usually some method of managing so that the amenities may still be enjoyed to some degree without substantial resource impairment. The question of degree in meeting goals is often a challenge to the professional, and a bit of imagination will lead to numerous possibilities for maintaining multiple resource values without costly compromises.

Regenerating stands, and plantations in particular, are high priority areas for weed control. Many states now have laws requiring regeneration of desirable species after harvesting; some require that a particular standard of success be met. In view of the strong tendency for cleared areas to "green up" quickly, there is usually a good chance a plantation will encounter weed problems. If the plantation is to succeed, weed control may be necessary, economically advantageous or legally required. The question is: How does one proceed?

There are many ways to weed a new plantation. One way is to avoid as much of the weed problem as possible by using appropriate logging methods and silvicultural practices. But many foresters inherit weed covered areas that must be treated, and an approach must be selected, or the resource abandoned.

Weeds may be treated in spots or broadcast, either before or after planting, with chemicals or with hand or mechanical equipment. The methods differ in safety, appearance, cost, effectiveness and biological consequences. Each method has been tested, and it is the forester's professional responsibility to know the consequences of any choice he might make among chemical and non-chemical methods. More knowledge may be required to employ herbicides than non-chemical treatments in terms of public inquiry. But biological effects are more complicated with the non-chemical methods because of physical as well as biological impacts, and the professional will need to understand the impacts of the various types of disturbance in some depth. The next few paragraphs provide some perspective of how approaches may be compared, even though our interest here is primarily with herbicides.

All vegetation control reduces plant competition. All methods of reducing plant cover cause localized ecological changes. That is their purpose. The various methods differ in their success in reducing competition, and in other impacts of their use. Each approach has certain unique features. Heavy machinery, such as bulldozers, simply cannot control vegetation without causing major physical disturbance. It disturbs soil, displaces virtually all non-target plants and animals, and may cause siltation in streams and interference with soil drainage. Broadcast application of herbicides has no direct physical impact, but leaves a chemical whose biological effects persist for some time after the helicopter had departed. Evaluation of total impact entails a careful examination of the unintended results as well as the intended effects. A good part of this evaluation depends on the duration of effects, and species affected.

The detailed effects of mechanical practices are beyond the scope of this manual. The persistence and behavior of the various herbicides are discussed in Chapter 5. It is significant that the unintended effects of herbicides are generally limited to fewer species and for shorter duration than those of other methods of weed control. This and cost advantages are the principal reasons why herbicides have found such extensive use in forests. Table 2-2 illustrates a generalized comparison of total impacts, and their importance to various resource values in long- and short-term comparisons. Impacts of successful planting without vegetation control are given to establish the consequences of re-establishing trees.

The above discussion of vegetation control approaches serves only to give a perspective to the potential herbicide user or how herbicides should be compared with other practices designated to achieve the same goal. The remainder of this chapter will be devoted exclusively to considerations of herbicides.

Herbicides and Forest Vegetation Management

Herbicides have several technical features of unique importance in implementing a vegetation management program. They are regulated by federal law; each herbicide has unique technical and biological properties; and their specificity offers considerable precision in prescription if operations are carefully planned. Consideration of the topics that follow will reduce errors in use, and improve net benefits after treatment. These topics may also form the basis for instruction of work crews and for discussions of herbicides with the general public.

Table 2-2. Comparative environmental impacts of forest practices on resource

Practice	Timber values	Water quality	Water yield
Tree planting[2]	Great (+) (if successful) Long	Small Long	Small Long
Vegetation control during forest establishment, by method and target species group			
Chemical:			
Herbs	Great (+) Long	Small (-) Short	Small (+) Short
Brush	Great (+) Long	Small (-) Short	Moderate (+) Short
Weed trees	Great (+) Long	Small Short	Small Short
Mechanical:			
Herbs	Moderate (+) Long	Moderate to great (-) Short	Moderate (+) Short
Brush	Great (+) Long	Moderate to great (-) Short	Moderate (+) Short
Weed trees	Great (+) Long	Small Short	Small Short
Controlled burning without chemicals			
Herbs	Small (+) Short	Small (-) Short	Small (+) Short
Brush	Moderate (+) Short	Small (-) Short	Moderate (+) Short
Weed trees	Moderate (+) Short	Small (-)	Moderate (+) Short
Hand clearing			
Herbs	Small (+) Short	Negl.	Negl.
Brush	Small (+) Short	Negl.	Small (+) Short
Weed trees	Moderate (+) Long	Negl.	Small (+) Short

[1]Plus or minus relates to a presumed value change relating to a specific resource.
[2]Impact of planting depends on success, which in turn may be determined by the effectiveness of cultural measures.

and amenity values, and their durations[1].

Big game	Small game	Aesthetics: Hiking	Aesthetics: Highway travel
Great (-) Long	Great (-) Long	Great (+) Long	Great (+ and -) Long
Moderate (-) or (+)[3] Short (+) Long (-)	Moderate (-) or (+)[3] Short and long	Small Short	Small Short
Moderate to great (+) Short	Variable Long	Great (-) Short	Great (-) Short
Small to moderate (+) Short	Small (+) Short	Great (-) Short	Great (-) Short
Great (+) Short	Variable to great (-) Short	Moderate to great (-) Short	Small to moderate Short
Variable (+ and -) Short	Great (-) Short	Attitude dependent Short	Attitude dependent Short
Moderate (+) Short	Small (+) Short	Small to moderate (-) Short	Small to moderate (-) Short
Great (+) Short	Great (+, -) Short	Small (-) Short	Small (-) Short
Great (+) Long	Great (-) Short	Great (-) Short	Great (-) Short
Moderate (+) Long	Great (-) Short	Great (-) Long	Great (-) Long
Negl.	Negl.	Negl.	Negl.
Moderate (-) Short	Small (+) Long	Great (-) Long	Great (-) Short
Great (-) Long	Great (+) Long	Great (-) Long	Great (-) Long

[3]Selectivity range of the herbicide will determine whether residual vegetation provides a more or less favorable habitat.

Labeling of Herbicide Products

Herbicides have certain unique properties that make them effective for economic control of plant pest problems. In order to be marketable, each product must meet certain criteria for registration. These criteria are very stringent, and a label is issued by federal or state authorities only after the manufacturer has provided thorough documentation that the herbicide meets both efficacy and safety standards when used as directed. The registration procedures are an excellent safeguard for both the users and for the general environment, and acute or chronic intoxication cases attributed to herbicides are rare or lacking where products have been used as directed.

The importance of the label as a legal document cannot be over-emphasized. The label must contain all pertinent information necessary for use, including full explanation of active ingredients, their concentrations and their formulations. The number of pounds of active ingredient per gallon is given for convenience in mixing large quantities, and full instructions are given for mixing any quantity for any use within the scope of the usefulness of the product.

Every product has several names. On a label, the prominent name is the *trade name*. Since several manufacturers may sell the same product under different trade names, the full *chemical name* is also given. Because this is often a jawbreaker that even chemists cannot utter with dignity, a *common name* is given too (Table 2-3). The common name is designated officially by the Nomenclature Committee of the Weed Science Society of America. Typical examples of the three types of names, all of the same material would be:

Table 2-3a. Common and product names for forestry herbicides.

Common Name	Other Names
Amitrole, Amitrol-T	Amino triazole, Cytrol
AMS	Ammate, ammonium sulfamate
Asulam	Asulox
Atrazine	Aatrex
Bifenox	Modown
Cacodylic acid	Silvisar 510, Phytar
DCPA	Dacthal
Dalapon	Dowpon, Dowpon M
Dicamba	Banvel-D
Dichlorprop	2,4-DP
Dinoseb	DNBP, Dinitro, Premerge
Diphenamid	Dymid; Enide
Fosamine	Krenite
Glyphosate	Roundup

Common Name	Other Names
Hexazinone	Velpar
Mineral spirits	Stoddard's Solvent
MSMA	Silvisar 550; Glow-on
Napropamide	Devrinol
Paraquat	Gramoxone
Picloram	Tordon; Amdon
Pronamide	Kerb
Silvex	2,4,5-TP, Kuron, Weedone
Simazine	Princep
Triclopyr	Garlon
2,4-D	Esteron, Weedone, many others
2,4,5-T	Esteron, Weedone, Vertac, many others

Table 2-3b. Examples of different names for herbicide products.

Trade Name	Common Name	Chemical Name
Esteron 99 Concentrate	2,4-D, LV Ester	Propylene glycol butyl ether esters of 2,4-dichloro-phenoxyacetic acid acid
Princep	Simazine	2-chloro-4,6-bis-(ethylamino) -s-triazine
Roundup	Glyphosate	N-phosphonomethylglycine isopropylamine salt

Precautionary statements are given on each label, but an exact measure of toxicity is not usually provided. Antidotes are described for products of substantial toxicity to humans. Instructions are usually given to keep the product out of reach of children, to mix and apply carefully, and to avoid contamination of water. It is good practice to observe such precautions. Since some herbicides are actually registered for application to water under some circumstances, and minor contamination may be inevitable with aerial applications, users should consult the section of this manual dealing with water quality. Regardless of the general properties of the product, the label is a document with the force of law. All printed material furnished by the manufacturer, and all advertising and technical information furnished by the manufacturer's technical representatives, are regarded as supplemental labeling and are legally binding on the manufacturer to be technically correct.

Properties of Herbicides Related to Effect

An herbicide is effective only if it enters into the metabolic processes of the target plant. Because most herbicides are selective in some or all uses, there is a need for them to be able to enter plants selectively, or to be able to exist in some plants without causing injury. The ability for a chemical to enter a plant and move within it is therefore basic to its effectiveness.

Herbicides enter plants in different ways. The route of entry is related to the physical properties of the chemical, the liquid carrier in which it is applied, the season of application and the characteristics of the plant. These determine whether the material enters leaves, stems or roots, and where it is concentrated within the plant. The chemicals also have biochemical properties that determine which metabolic processes within the plant are affected once the chemical has reached its destination.

Herbicides that enter plants through foliage must be physically able to penetrate leaf membranes and cuticle, and to be able to diffuse through stomata. Such chemicals must be soluble in waxes or oils to penetrate cuticle, but water-soluble substances will penetrate cell membranes if not covered by heavy cuticle. Thus, condition of foliage may determine which properties are most desirable.

Applications to bark must be oily if they are to penetrate effectively without being injected into a wound. Oil solutions are more effective than emulsions on bark. If a wound is created, as in injection, the chemical is placed directly in the vascular system, where water solubility is necessary.

Roots absorb chemicals that enter the soil solution. An effective soil-active chemical (also known as "residual" herbicide) must be able to enter the soil by rainfall action, and to remain available in soil for some time without leaching into water supplies. They may vary in solubility, but must remain held in soil by physical adsorption, during which gradual desorption leads to low concentrations in the soil solution. As long as this process occurs in surface soils, plants gradually absorb the trace amounts in the soil solution, and eventually absorb enough for control.

Parent herbicidal compounds are often physically incompatible with economic use. They must therefore be modified and prepared as products with desirable properties.

Formulation

Physical properties of herbicides are determined either by the nature of the active component of the product, or by some reaction product that may alter its original properties. Among the chemicals applied as the parent compound are amitrole, atrazine, cacodylic acid, dinoseb, fosamine pronamide, simazine, and hexazinone. The other herbicides are usually salts or esters of the parent compound.

As shown in Table 2-4 herbicides are sold as water soluble

Table 2-4. Herbicides registered for use in forestry in the USA (1980) and properties of products made from them.

	Formulations[1]								
Common Name	GF	WSP	Granular	Pellet	WP	OSL	WSL	FC	EC
Asulam*							X		
Amitrole*					X		X	X	
Atrazine*									
Cacodylic acid							X		
Dalapon*		X							
Dichlorprop*									X
Dicamba*						X	X		X
Dinoseb***						X	X		X
Fosamine ammonium***							X		
Glyphosate***							X		
Hexazinone***		X		X					
Picloram*			X	X		X	X		
Pronamide***					X				
Silvex*†									X
Simazine*			X		X			X	
2,4-D*						X	X		X
2,4,5-T*†						X	X		X
Triclopyr						X	X		X
MSMA							X		
Nursery & ornamental use only									
Bifenox**					X			X	
Diphenamid					X				
DCPA					X				
Paraquat							X		
Napropamide**									
Chloropicrin	X								
Methyl bromide	X								

* Registered for aerial spray.
** State registration, check local regulations.
*** State-registered for aerial application, check local regulations.
† Suspended for forestry uses by the EPA pending outcome of cancellation proceedings.

[1]GF = Gaseous fumigant (high pressure).
WSP = Water soluble powder
Granular = 1-20 percent granular
Pellet = 5-10 percent pellets
WP = 25-80 percent wettable powder
OSL = Oil soluble liquid (non-emulsifiable in water)
WSL = Water-soluble liquid
FC = Flowable concentrate (liquid suspension of solid)
EC = Emulsifiable concentrate (non-water-soluble)

powders, granules, pellets, wettable powders, water soluble liquids, emulsifiable concentrates, or as flowable liquids. Commercial products also include "inert" additives, such as emulsifiers, surfactants (wetting agents), cosolvents and stabilizers. Some are merely concentrated solutions of solids in water or other solvents. In each product, the parent compound is prepared in a form in which it can be applied for a particular use without modifying its biological properties. It is useful to know how the product properties affect use.

Water-Soluble Liquids

When a product is sold as a water soluble liquid prepared as a salt, the user can be assured that the compound is not likely to precipitate in a water spray mixture. Water soluble acids are sometimes reacted with a base to form a neutral salt. Examples are monosodium methanearsonate (MSMA), a reaction product of methane-arsonic acid with sodium hydroxide, and neutralized cacodylic acid. The use of neutral salts reduces corrosiveness to equipment.

Water-soluble liquids are also prepared by reacting insoluble products such as phenoxy herbicides, dicamba, picloram, asulam, and ghyphosate with bases, usually amines, to form a salt. Unlike the products above that are soluble when unreacted, these materials can become insoluble and precipitate if introduced into an acid medium. It is very important that they be kept away from mixtures with acidic components. In all other respects, they are handled as if they were completely water soluble.

Water-soluble compounds may be used as foliage sprays, as injectables or as soil-active herbicides. They do not penetrate bark efficiently, and they are relatively inefficient at penetrating waxy leaves unless fortified with surfactants. Without surfactants, water soluble chemicals have a tendency to form beads of moisture rather than a thin film on foliage, and absorption may be poor. This is a useful property when foliage is young and tender, and where too-rapid penetration will cause damage to translocating tissue before sufficient active chemical is moved into the stems. It reduces effectiveness on mature foliage.

Injectable concentrates must be water soluble to move efficiently through the vascular system of a tree. Ester preparations or other soil-soluble mixtures are inefficient in a tree wound, although they will provide control when applied in excess.

Water soluble herbicides are usually heavier than water. Although completely soluble, they need to be mixed when preparing a spray. If not agitated at some point, the heavy concentrate may form a layer on the bottom of the tank. This will lead to incorrect application rates even though the equipment is properly calibrated and the correct amount of herbicide is used. This is especially important when mixing in a helicopter unless it has vigorous by-pass agitation.

Water Soluble Powders

Water soluble powders are limited to amitrole and dalapon. Amitrole is soluble as the parent compound, dalapon is a salt. Amitrole is also prepared as a liquid, Amitrol-T. Hexazinone is of limited solubility, and solubility varies with temperature. It is not a salt, and solubility cannot be improved without added solvents. In dilute solutions it behaves as a water-soluble material, but at concentrations used for aerial applications, its limited solubility can cause the formation of crystals. It is important to follow label instructions to avoid plugging equipment.

Once mixed, water solutions do not settle out unless crystallization occurs in a super-saturated solution.

Wettable powders, flowable concentrates, granular and pelletized products

Wettable powders, flowable concentrates, granular and pelletized products are materials normally designated for application to soil. Some of the powders and flowables have some capacity for absorption into foliage, as well. Wettable powders are products made from herbicides of low solubility in water. They are normally prepared as finely ground dusts, mixed with attapulgite clay and perhaps other dispersing materials. They are nearly always applied in water, but they may be dispersed in oil or mixtures of water and oil under some circumstances. Under no circumstances do the wettable powder spray preparations become solutions beyond the very limited solubility of the herbicides. The trade name will often designate the concentration of active ingredient and type of formulation. Thus, a product name followed by 80W is a wettable powder containing 80 percent active ingredient, by weight.

Flowable concentrates are very finely ground wettable powders suspended in a liquid matrix for ease of handling. These materials are stabilized in the liquid product by the addition of gelatinous stabilizers or similar materials. These additives will not enhance the stability of spray preparations significantly after mixing because the stabilizer may not be adequate when diluted. As a general principle, all wettable powders and flowables should be applied with agitation in the sprayer. Mechanical agitation is usually preferable to liquid by-pass circulation.

Flowable concentrates also frequently have the concentration given on the product name. A flowable may have a 4L after its trade name, signifying that it contains four pounds per gallon, active ingredient (a.i.), as a liquid. Because there is some inconsistency in trade name nomenclature, it is wise to check the label statement of pounds per gallon or percent active ingredient for flowables as well as other products.

Flowables are slightly more costly than wettable powders, but are equal in effectiveness per unit of active ingredient. Flowables are easier to handle, especially in small equipment. The liquid preparation also eliminates dust inhalation.

Granular and pelletized products are prepared for easy distribution as dry materials. Only picloram, simazine and hexazinone are presently prepared in these forms. Granular products are normally bound to sand particles in concentrations suitable for distribution with fertilizer spreaders or similar equipment. Pelletized materials are usually applied by hand in spot applications, but also may be used broadcast. Pelletized and granular materials must be kept dry, or they become unmanageable.

Oil-soluble liquids and emulsifiable concentrates

Oil-soluble liquids and emulsifiable concentrates are made from acids that have been reacted with alcohols or organic bases to form oil-soluble esters or amines. Phenoxy herbicides, picloram and dicamba may be prepared in these forms; most phenoxys are used in ester form. A few products are simply made with parent acids dissolved in organic solvents with emulsifiers. These are known as "solubilized" acids, which are usually applied as oil solutions or water-oil emulsions.

Oil-soluble amines and esters have several properties that are important in use. They are soluble in oil but not in water. The oil solubility helps them penetrate bark and waxy foliage. They may be mixed in water when emulsification agents are incorporated in the formulation to disperse oil droplets in the water. Emulsification produces a milky mixture. Emulsions must not be confused with solutions. If a mixture with water is a clear liquid, it is a water-soluble product. If it separates into a two-phase mixture with the water on top, it is a *non-emulsifiable,* oil-soluble concentrate. If it forms a milky emulsion, it is an *emulsifiable* concentrate. Most esters are sold as emulsifiable concentrates.

Certain products sold for application in oil as aerial sprays are non-emulsifiable. Users should not attempt to use them in water because they will not form an even mixture. Sometimes emulsifiers break down in storage, with the result that the product becomes non-emulsifiable, and must be applied in oil.

Esters and oil-soluble amines are normally more expensive than water-soluble amines and other salts. They are used because of their strong affinity for leaf surfaces and ability to soak through bark. They do not wash off foliage readily, and do not move in soil. For aerial application, esters are more effective than water-soluble formulations of phenoxy herbicides. Picloram, however, appears more effective as a water-soluble material except when applied to the outside of bark.

Esters have a serious disadvantage in some situations. They have some ability to evaporate after application, and the vapors may drift and cause damage to nearby crops. Esters vary in their volatility. Esters made from alcohols with four carbons or less are known as "high-volatility" esters. Those with six or more carbons in the alcohol are "low-volatility" materials. Oil-soluble amines have no volatility. All oil-soluble herbicides registered for use in forest management at present (including emulsifiable products) are low-volatile esters. Used with proper precautions, drift should be minimal and crop damage should not be a significant problem (see Chapter 4).

Unformulated Products

A number of compounds have properties that render them suitable for use as herbicides without being formulated. Many of these are solids of low solubility in water, ground to a fine texture and sold as wettable powders with 50 percent or more active ingredients. Others are sold as water-soluble powders ready for mixing. They have no surfactants, and they have not been modified chemically.

Gaseous Products

Soil fumigants are a special kind of soil-active herbicide. The gaseous fumigant methyl bromide (containing chloropicrin as an odor marker as well as biocide), is sold only in pressurized containers. High vapor pressure of fumigants dictates that they be injected into the soil and covered with an impervious layer, such as polyethylene. They are hazardous to use for this reason, and must be used by certified applicators with appropriate equipment. Fumigants are used as herbicides only in nurseries.

Matching Formulations to Use

The use pattern of a chemical determines the choice of formulation. Esters in oil are best suited for bark sprays. They also kill foliage, but too quickly for best results on new foliage.

Amines and other liquid water-soluble products are primarily useful for foliage sprays in water, or for injection without substantial dilution. Wettable powders, flowables, granulars and pellets are usually applied as soil-active herbicides. It is unwise, and often ineffectual or illegal, to use a product for a purpose that is not on the label, even though other products made from the same active ingredient may be formulated for that use. The tendency to use a lower-priced product with a similar trade name is not necessarily appropriate. Several products may have the same generic name (e.g. Weedone, Esteron, etc.), but may actually contain completely different active ingredients. It is wise to become familiar with the properties of the active ingredients, then shop for the formulation and product that is best suited for the job.

Biological Properties and Uses

The various families of herbicidal compounds have typical patterns of activity in plants. For some, the specific biological mechanism of action is not know, but for the most important groups used in forestry, the important mechanisms are reasonably well understood.

Phenoxy herbicides

Phenoxy herbicides include 2,4-D, 2,4,5-T, dichlorprop and silvex. They are used for control of broadleaf weeds and woody vegetation. These materials differ little from each other chemically, but act somewhat differently as herbicides. They are not simply different formulas of 2,4-D, for example. The diagrams shown here illustrate the chemical differences.

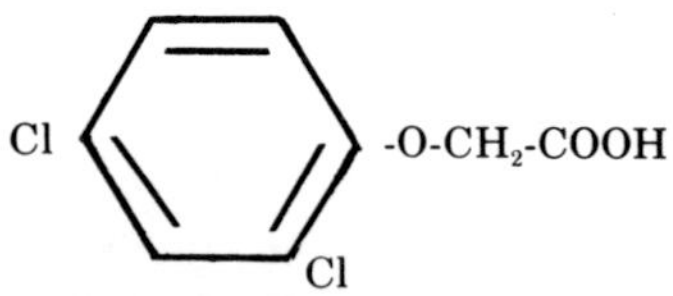

2,4-Dichlorophenoxyacetic acid (2,4-D)

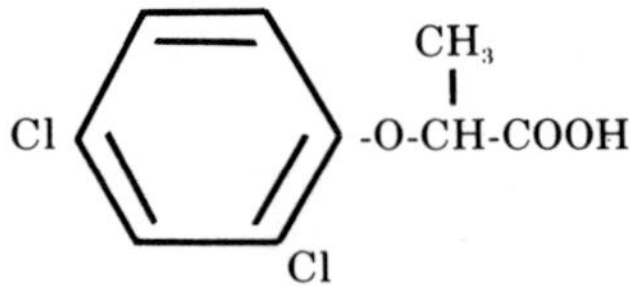

2-(2,4-Dichlorophenoxy)-propionic acid (Dichloroprop)

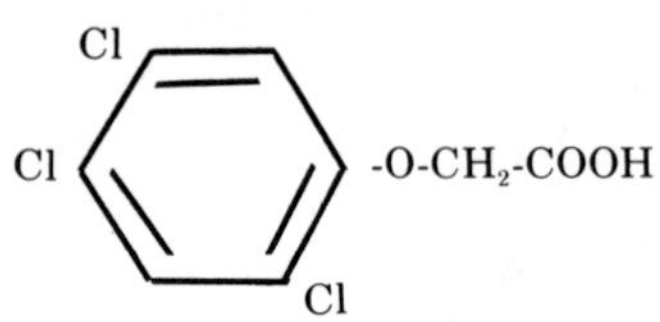

2,4,5-Trichlorophenoxyacetic acid (2,4,5-T)

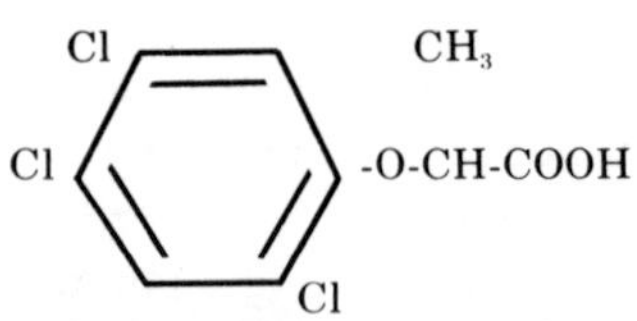

2-(2,4,5-Trichlorophenoxy)-propionic acid (Silvex)

The difference between 2,4-D and 2,4,5-T is simply an extra chlorine atom on the benzene ring. The difference between 2,4-D and dichlorprop is the addition of a methyl group on the alkanoic acid side chain. Silvex (2,4,5-TP) is to 2,4,5-T exactly as dichlorprop (2,4-DP) is to 2,4-D. These minor substitutions influence the accessibility of various parts of the molecules to metabolic action because they must fit closely to particular enzymes in order to produce their effects. If the fit is poor, the effect is slight.

All phenoxy herbicides are known as synthetic growth regulators. This is because they interfere with cell wall processes, especially elongation, and typical symptoms of damage include curling of foliage when one side of an elongating twig is affected more than the other. The asymmetric growth is actually the result of disruption of more, basic processes in which the herbicide affects the respiration system, and perhaps the RNA transfer system during cell elongation and division. Cell division may proliferate; rapid division is not associated with irreversible genetic changes.

The phenoxys are translocated herbicides. This means that they are absorbed from foliage or bark, and moved somewhere to a specific site of activity, which is normally in the growing tips, with some deposited in the cambium, roots and storage tissues. Because of this, it is important to apply the herbicide in a way that avoids killing the translocating tissue before the herbicide has moved to the site of action in sufficient quantity. Chemical is normally moved away from foliage through the phloem or inner bark toward some growing tip or cambial growth site. Application in overly high concentrations, or with a surfactant that causes too rapid invasion of tender tissues with chemical, will cause rapid brownout and phloem collapse. This causes poor ultimate translocation of chemical, with resulting poor control of stems and roots, and extensive sprouting. The appearance of immediate success is misleading. One of the reasons for using 2,4,5-T instead of the lower-cost 2,4-D is that it has less tendency to "burn off" translocating tissues, and thus results in better control of many woody species. Where 2,4-D absorbs and translocates effectively it is often a useful chemical for woody plant control. Only in very limited circumstances does 2,4-D add to the effectiveness of 2,4,5-T on 2,4-D resistant species. Inappropriate use of 2,4-D will cause injury to conifers, and only temporarily set back the brush, thereby increasing the difficulty of control later.

The ability to cause foliage curling at extremely low concentrations makes phenoxy herbicides, especially 2,4-D, the group most likely to produce visible injury to certain non-target plants. In addition, their preparation as esters with some volatility enhances their ability to move off target. For this reason, phenoxys must be used with great care within a mile of commercial crops or ornamental plants of major sensitivity. These include most of the broadleaf crops, such as beans, tobacco, cotton, grapes, potatoes, tomatoes and peanuts, to name the most sensitive. None of the grasses or grains are sensitive to such drift and many of the root crops and fruit and nut trees are intermediate in sensitivity. Most of the phenoxy herbicides used in the United States are applied to grain crops for control of broadleaf weeds without injuring the crop.

Biological activity of any herbicide varies with season. The phenoxys are most active when the target species have fully mature foliage that is exporting carbohydrates to roots and stem tips. If applied earlier than this, rapid desiccation of foliage occurs and the plants sprout vigorously. Applied later, foliage has accumulated waxy cuticle on its surface that restricts uptake and translocation. Moreover, growing tips may have become dormant and no longer the recipient organ for most translocated herbicide. The effect of bark application varies less with season, because absorbed herbicide will cause extensive girdling and systemic action, regardless of season, when applied as a basal spray. When applied as an aerial spray, there is not enough fluid to cause diffusion of the chemical through thick bark, and only aerial portions with thin bark may show damage.

Phenoxy herbicides do not damage conifers during certain dormant seasons. The short-needle conifers with discrete buds are normally dormant enough to spray without injury from bud maturity in the fall until budburst in spring. This permits a spring or fall dormant spray for release. Severe injury will result from application during the period of rapid elongation in late spring. Western pines become tolerant at about the same time in the fall, but lose their tolerance earlier in the spring. For example, ponderosa pine will be severely injured if phenoxys are applied in March, even before signs of growth have started; early February is about the end of its resistant period. Southern pines have considerable tolerance while elongating in late June and July. Spruces appear the most tolerant group of conifers, and an early summer application may selectively increase the proportion of spruce in a mixed conifer stand.

In order of usefulness in forest applications, 2,4,5-T is by far the most versatile of the phenoxys. Among the phenoxys, 2,4,5-T normally has the least effect on conifers, and has the broadest spectrum of activity on hardwoods and shrubs. 2,4-D may be mixed with 2,4,5-T when 2,4-D sensitive brush species are abundant. Silvex and dichlorprop are not adequate substitutes for 2,4,5-T and 2,4-D. The prevailing tendency to use 2,4-D with 2,4,5-T is usually poor strategy. Experiments have shown consistently that 2,4,5-T produces most of the desirable silvicultural effects on 2,4-D resistant species. Species sensitive to both 2,4-D and 2,4,5-T are treated with 2,4-D alone. Both herbicides are very important in complementary roles.

Silvex is used on a limited basis for control of certain oaks and maples. It has been used as a substitute for 2,4,5-T because of presumed toxicological problems with 2,4,5-T. This is inappropriate, because the toxicological properties are almost identical, and silvex is less desirable in most silvicultural operations.

Dichlorprop is marginally more effective than 2-4-D on certain species.

The phenoxy herbicides, and especially 2,4,5-T and silvex, have come under attack in recent years because of the use of herbicides in military action, and because of the discovery of a dioxin contaminant (TCDD) in 2,4,5-T and silvex. The phenoxys and TCDD have been the subject of an immense amount of research in the past decade, and are presently better understood than virtually any other pesticide or toxicant. The evidence from laboratory animals, applicators and persons exposed in industrial accidents indicates that there is a substantial margin of safety in humans heavily exposed to these materials. At this time, the Environmental Protection Agency is holding cancellation proceedings against 2,4,5-T and silvex on the presumption that the safety margin is not great enough.

Triazine herbicides

Triazine herbicides include the symmetrical atrazine and simazine, and asymmetric hexazinone. These are characterized by six-memberd rings containing three nitrogen and three carbon atoms. They vary in solubility, with simazine being very insoluble, atrazine having very limited solubility and hexazinone being quite soluble as soil residual herbicides go. Their effectiveness as foliage sprays is closely related to their solubility.

Atrazine and simazine are generally used for control of germinating or early growth of herbaceous weeds in coniferous plantations. Hexazinone may also be used on advanced weed

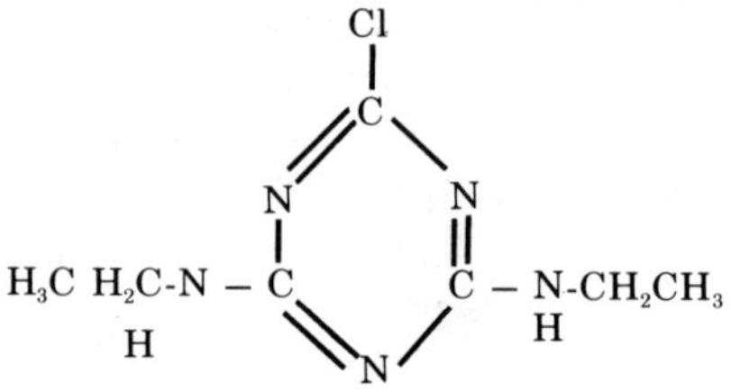

Simazine
2-chloro-4,6-bis-
(ethylamino)-s-triazine

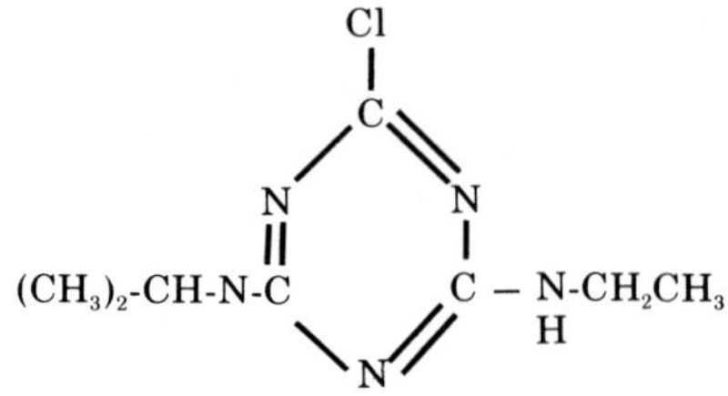

Atrazine
2-chloro-4-ethylamino-
6-isopropylamino-s-triazine

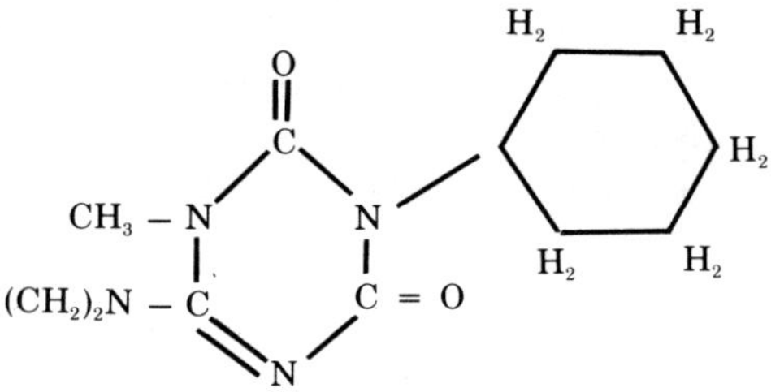

Hexazinone
3-cyclohexyl-6-(dimethylamino)-1-
methy1-1,3,5-triazine-2,4(1H,3H)dione

growth, and in some instances on brush. Their biological activity is greatest on grasses, but extends to some broadleaf weeds; at high rates of application, most herbs are affected. Their limited penetration into soil gives planted conifers some protection, and they are generally not affected. This physical selectivity is lost with small container seedlings, however, and some injury may result.

The biological effects of the *s*-triazines appear to be largely confined to photosynthetic processes. The Hill reaction, in particular, is inhibited, and causes starvation and photolytic destruction of chloroplasts, leading to inability to recover. Nitrogen metabolism is also affected, possibly leading to excessive concentrations of ammonium ion. Hexazinone is still under study, but appears to cause the same general range of symptoms, which extend to some hardwood species. The greater solubility of hexazinone adapts it to some uses to which *s*-triazines are unsuited, and leads to risk of conifer damage on light-textured soils.

Unlike the phenoxys, some species are biochemically tolerant of the triazines through selective detoxication. Members of the corn family dechlorinate atrazine or simazine to hydroxy compounds with no biological activity. They cannot convert hexazinone, presumably because of a different structure (methyl group) at the key reaction site. Conifers appear generally to enter into a chemical reaction with *s*-triazines known as conjugation, leading to biologically inactive combinations with plant proteins. For this reason, conifers appear able to tolerate moderately large amounts of the herbicide. It is possible to cause injury with overdosages, however, because the conjugation process is relatively slow. Metabolism of hexazinone is currently being investigated.

The phosphonoalkyl compounds

The phosphonoalkyl compounds include fosamine and glyphosate, which may be used for selective brush control. These newcomers are less well understood than the phenoxys and triazines, but they do have certain clear patterns of action. These are highly translocated herbicides that do very little damage to translocating tissues. They are thus relatively intact when they arrive at their sites of activity. They have no soil or root uptake activity, and enter entirely through foliage absorption and translocation.

The phosphonalkyl compounds have the following structures:

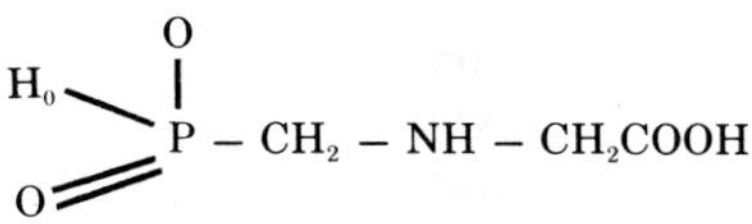

Glyphosate
N-phosphonomethyl-glycine

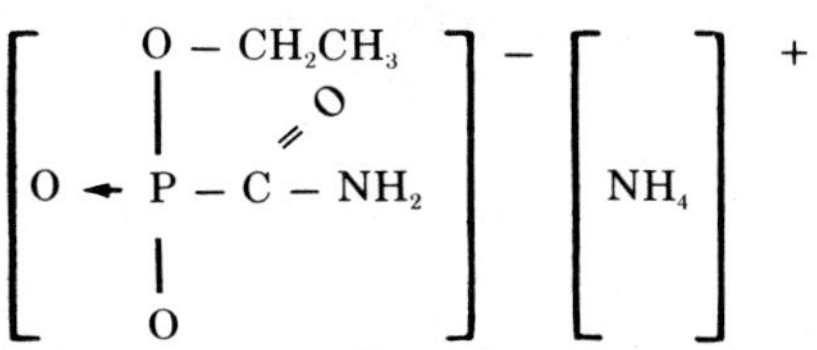

Fosamine
Ammonium ethyl carbamoyl phosphonate

Glyphosate, in particular, has been shown to have a specific metabolic role in inhibiting the formation of an essential amino acid in plants, phenylalanine. Because this substance is crucial for survival, inhibition of synthesis leads to metabolic failure and death of many plant species. Phenylalanine is not synthesized by animals and thus glyphosate is virtually non-toxic to animals. Glyphosate is effective on most herbs and deciduous plants and trees. It is especially effective on perennials when applied in fall so as to translocate into the roots just before leaf fall.

Glyphosate is metabolized in the light, in foliage. One hypothesis suggests that this is a major reason for maximum effectiveness on deciduous species in fall; the metabolizing organs are absent in winter, permitting a long internal exposure to the chemical without metabolic depletion of the chemical. The resistant species, such as conifers and evergreen hardwoods and shrubs, carry foliage through the winter and emerge without injury in the spring, presumably having deactivated the chemical. This hypothesis has not been fully verified. Even among the deciduous species, there is some variation in sensitivity, and selectivity may be observed among hardwoods at low rates. Sugar maple, for example, is quite tolerant of glyphosate. The mechanism for such tolerance is not known.

Fosamine behaves similarly to glyphosate in some respects. The specific site of action appears to be the elongating, or "intercalary," meristems, where it inhibits elongation. Its seasonal selectivity is nearly identical to that of glyphosate but it is far more limited in the number of species it controls.

The herbicidal effects of phosphonalkyl compounds are reduced if vegetation is disturbed prior to application. Even minor injury, such as browsing, will apparently reduce activity significantly. For maximum effectiveness, the materials are not recommended for use within one year after treatment with phenoxys or mechanical disturbances such as logging or scarification.

Organic arsenicals

The organic arsenicals are used in forests only as injected herbicides. These materials are different from the inorganic arsenicals (arsenites) frequently mentioned in murder mysteries. The principal difference is that the highly toxic inorganic arsenic preparations are highly reactive and unstable. The organic arsenicals are more stable and are far less toxic. There is still some speculation as to the specific sites and modes of action in all organisms, although there is evidence that the arsenicals substitute for phosphorus in metabolism, leading to plant respiratory failure. There is no clear explanation as to why the organic arsenicals are specific plant toxins, whereas the inorganic forms are general toxicants.

The arsenicals are metallic compounds. They are thus presumed to be ultimately persistent in the environment. This has recently been proved wrong when it was discovered that the herbicidal arsenicals are naturally occurring substances, some of which are metabolized to volatile arsines. In one of nature's neatest tricks, the arsenicals move from form to form and place to place in a global arsenic redistribution system, in which some forms can be herbicides and at least one appears to be a useful and highly transient gaseous insecticide, synthesized by the molds brought in by the insects!

$$As_2O_5 \longleftrightarrow AsO_4 \longleftrightarrow CH_3\ AsO(OH)_2 \longleftrightarrow (CH_3)_2AsO(OH) \longleftrightarrow (CH_3)_3As$$

Arsonate (oxidized)	Arsenite	Methanearsonic acid (MAA)	Cacodylic acid	Trimethylarsine gas (Reduced)

In this reaction sequence, various types of organisms can convert the specific materials to more oxidized or reduced forms, with the end result of creating a full array of substances in an array of concentrations at equilibrium. Given a suitable energy source, such as vegetation, there is a tendency toward reduction with the slow formation of trimethylarsine, which is a toxic and unstable gas. Most of the global arsenic, however, remains relatively inactive biologically, being in the oxidized form.

There is some evidence that the use of the sodium salt of MAA, (MSMA) and cacodylic acid not only kills conifers and hardwoods by injection, but produces arsine metabolite products. These are thought to be accountable in part for the control of bark beetles fumigated in their galleries in treated areas (see Chapter 3). There is no danger of intoxication of human, however, because toxic concentrations would be reached only under the bark. Ventilation would immediately dilute the arsines to harmless concentrations.

Picloram and dicamba

Picloram and dicamba are synthetic growth regulator-type herbicides, similar in effect to phenoxys, but unrelated chemically. Picloram and dicamba are persistent for up to a year (and occasionally longer), and have biological activity through root and foliage uptake. Neither is particularly selective among woody species, but both are ineffective on grasses. Although their specific metabolic actions are not known in detail, they cause leaf curling, phloem collapse and systemic damage. They are used because their persistence can contribute to a long-term treatment of resistant woody species, hampering recovery of susceptible plants. Both damage conifers severely if applied to foliage or roots. For this reason their use is limited to site preparation or tree injection.

Picloram is highly mobile in root systems. Shrubs or trees dying from picloram injury may release picloram unaltered from roots in quantities great enough to injure nearby trees, a phenomenon called "flashback." Heavy dosages should not be used where conifers can pick up the residues, and spills or dripping should be avoided. Picloram is also extremely toxic to potatoes, tobacco, peanuts and certain other crops. It is extremely difficult to remove entirely from equipment.

Triclopyr

Triclopyr is a synthetic growth-regulator-type herbicide intermediate in many effects between picloram and 2,4,5-T. It is effective at certain seasons on broadleaf woody species, primarily through foliage uptake. Persistence is slightly greater than for 2,4,5-T. Metabolic action is not generally known, but is thought to be similar to picloram, despite a much lower degree of injury to conifers. Triclopyr is produced in both water and oil-soluble formulations, of which the water soluble amines are very effective for tree injection and for release of northeastern conifers. The esters are more effective, in general, as foliage treatments, but have less conifer selectivity.

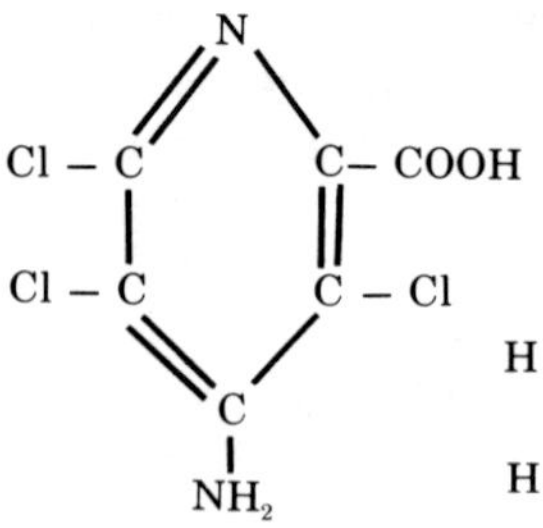

Picloram
4-amino-3,5,6-
trichloropicolinic acid

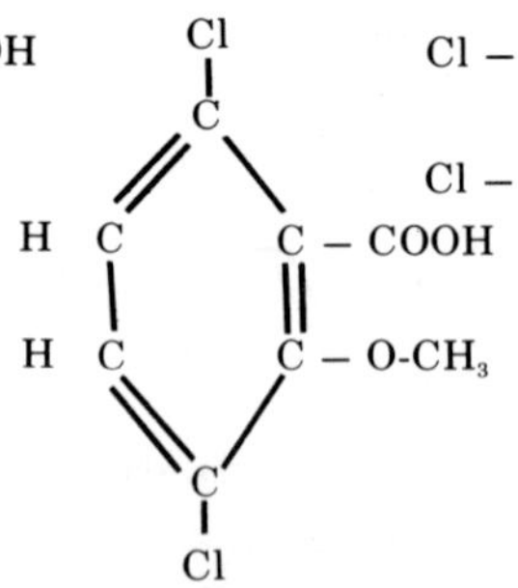

Triclopyr
3,5,6-Trichloro-
pyridinyloxyacetic
acid

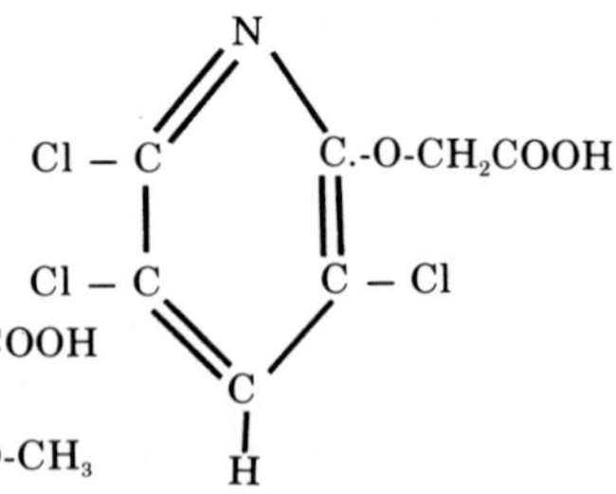

Dicamba
3,6-dichloro-
anisic acid

Asulam and amitrole

Asulam and amitrole are unrelated minor use materials that have a few properties in common. Both have some ability to inhibit synthesis of chlorophyll in new growth, and cause whitening of newly emerging foliage. Both inhibit elongation. Both materials are highly translocated, and depend on foliage uptake for herbicidal activity. Because of their lack of injury to foliage tissue, both must be applied relatively late in the maturation of foliage in order to insure translocation to roots.

Amitrole is a general herbicide used for control of herbs and woody species, and in limited circumstances has some selectivity for conifers. As a general matter, it kills by interruption of photosynthesis. Asulam, in contrast, has important herbicidal effects on bracken fern almost exclusively. After treatment, there are few signs of injury until the following year, when it restricts the ability of the rhizome to elongate into new fronds.

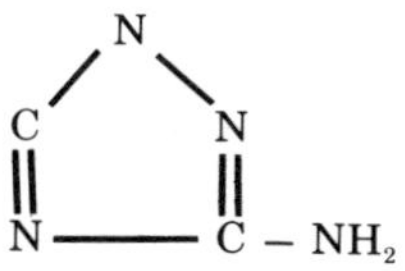

Amitrole
3-amino-s-triazole

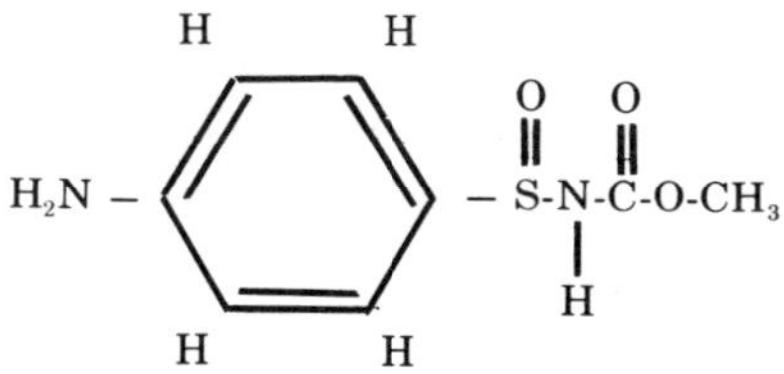

Asulam
Methyl sulfanilycarbamate

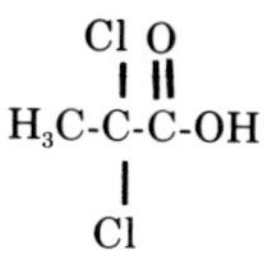

Dalapon
2,2-dichloropropionic acid

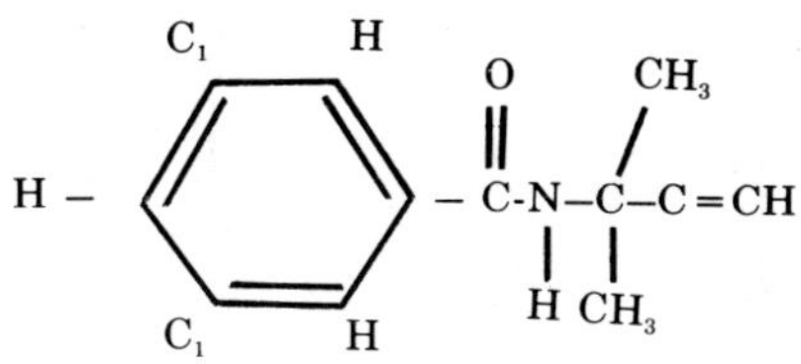

Pronamide
N-(1,1-dimethylpropynyl)-3,5-dichlorobenzamide

Dalapon and pronamide

Dalapon and pronamide are also minor use compounds that control grass. Dalapon is a chlorinated aliphatic acid that has activity through both foliage and soil. It appears to be active in the terminal respiration process in plants, but it is not known why its effects are limited to green plants, and specifically grasses. High rates of application can injure conifers, but for some reason still not explained, the addition of atrazine to the mixture can reduce this injury while increasing grass and broadleaf weed control. Dalapon is applied in spring, has short residual effect in soil, so conifers can be planted within two weeks. When mixed with atrazine, dalapon may be applied directly over some conifers. Occasional conifers will show symptoms of unseparated needles or slightly reduced growth. The effects are temporary. Some pines are sensitive to this mixture.

Pronamide is a carbamate herbicide used in some states specifically for control of perennial grasses in conifer plantations. Its mode of action in grasses is not well known, but its persistence and selectivity are comparable to those of atrazine. It is applied in fall, either before or after conifers are planted.

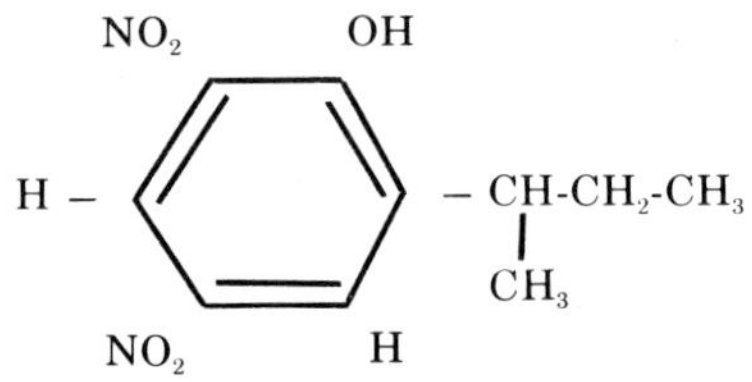

Dinoseb
2-sec-butyl-4,6-dinitrophenol

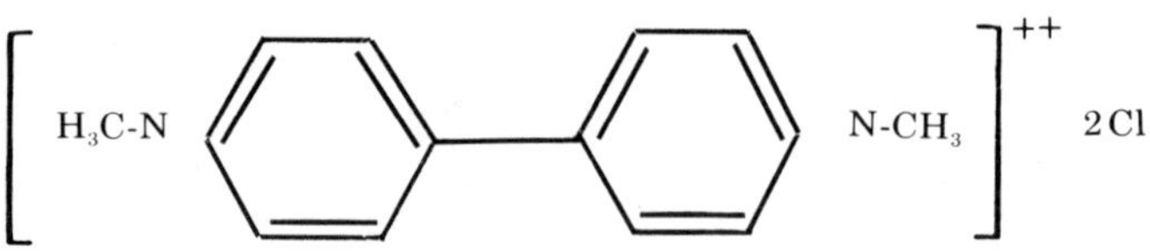

Paraquat
1,1-dimethyl-4,4'-bipyridinium
ion (as dichloride salt)

Dinoseb

Dinoseb is a foliage desiccant of limited use in forestry. This material is unlike any of the other forestry herbicides in that it is a *general toxicant.* This material has been used as an insecticide, fungicide and herbicide. It is apparently a general respiration inhibitor, and relatively low concentrations affect a wide variety of biota. For this reason, it is used in forests only as a pre-burn desiccant to prepare fuels a few days before controlled burns. It is the most toxic of the herbicides used in forestry, and should be handled with suitable caution.

Dinoseb has a very rapid effect on foliage. The quick brown-off prevents translocation of the herbicide, and prevents systemic injury. Although this is satisfactory for fuel preparation, it does not control sprouting species.

Paraquat

Paraquat is a bipyridinium ion compound with limited use as a nonselective desiccant in nurseries. Paraquat has special health considerations for applicators, and they should use it with suitable precautions.

Nursery residual herbicides

Nursery residual herbicides, including bifenox, diphenamid, DCPA and napropamide, do not have well known metabolic actions in plants. All have limited mobility and are very useful for inhibiting germination or emergence of herbaceous weed seeds in conifer seedbeds. Of these, only bifenox has appreciable ability to injure nursery weeds after they have emerged from the ground, i.e. "postemergence." Thus, preemergence timing is an important consideration for econonomic use.

Most of the activity of these herbicides is expressed as inhibition of the root radical. Unable to develop a root system, the weed fails to develop. Although most of these herbicides do not control visible weeds effectively, they may be used to prevent further emergence while hand weeding is in progress.

Conifers can tolerate napropamide, diphenamid and bifenox under certain conditions if applied before the conifer seeds germinate. A new seedbed may be treated immediately after sowing, and be kept virtually clean of weed germinants. DCPA can be used after conifers are several weeks old if the visible weeds are removed first. Mineral spirits (Stoddard's Solvent) can be used for control of small weed seedlings in preparation for DCPA. Label instructions must be followed carefully to avoid injuring conifers.

Soil fumigants

Soil fumigants are general biocides used for simultaneous control of weeds, insects, and particularly fungi in nursery beds. Methyl bromide with chloropicrin are nearly always applied together with a soil injection device that places the gaseous fumigant beneath the surface of soil, then covers the site with a plastic tarpaulin. The sealed soil retains the gas, which slowly diffuses through seedcoats and various other semi-permeable surfaces, causing removal of nearly all active organisms. Numerous spores remain, however, and disease organisms are normally among the resistant cultures. Because these materials injure desirable fungi, their primary use as weed killers is seldom justified. Special equipment is needed for this work.

Comparing the Effects of Herbicides

Among the herbicides described above, it will be observed that each herbicide may have different effects, depending on dosage, season applied and susceptibility of target and non-target species. None of the herbicides does just one thing, and there is some overlap among the things they do. The exact nature of the overlap is determined by the way they are used. Because none is exactly like any other, none is a complete substitute for another.

Foresters often gain experience with herbicides operationally to evaluate their effects in their own silvicultural situation. When trying out a herbicide for the first time, it is important to know beforehand when it should be applied, and how much is needed to control certain species. An unfamiliar chemical may well have a different *seasonal* optimum from a more familiar material. Tests should include several dates of application to allow the best timing to show up. The season for best results may be quite short. When comparing chemicals, each must be applied at its own best timing for a satisfactory comparison. Many good products are rejected because of inadequate testing in this regard. Only manufacturers representatives or scientists from bonafide research organizations may test chemicals that are not registered for the designated use.

Silvicultural Use of Herbicides

Proper prescription of herbicides requires a plan. Since several important principles are involved, we recommend a written plan to insure that all factors are included and each step carried out.

The first principle of prescription is that it must lead to achievement of objectives in concert with other operations. For example, if the goal is the maximum growth of softwood sawlogs, the full prescription entails provision for site preparation, establishment of conifers and stand tending. Perhaps the herbicide will only be used in the site preparation or release phases, but the benefits of spraying will be realized only through complete development of the crop. It is not what is killed that is important; reforestation and timber growth are the real objectives. Use of herbicides is only one step in that process. It must be compatible with all other parts of the plan, and it must be necessary.

The second principle is that the specific objective of vegetation control must be defined, and method selection tied to it. Target species should be identified as only those that are interfering with the general management objectives. The decision to use herbicides should be made with full consideration of alternative methods and their full range of impacts. The choice of method and/or chemical represents a weighed evaluation of costs, effacacy and impacts of all alternatives on a comparable basis.

The third principle is correct matching of herbicide, formulation, timing, dosage and application method, crop and target species sensitivity. Each factor requires careful attention, because an error at one point will often cancel out all other inputs of the operation.

The process of prescribing any pesticide use is a complicated one. The array of herbicidal products and target species is complex and few people will normally be involved in the prescription process. Laws and availability of products change constantly. Thus individual responsibility is heavy. Under the circumstances, it is desirable in any large organization to have at least one forester who is responsible for up-to-date knowledge of the subject, and for any user to be current on the particular use contemplated. Smaller growers should consult qualified foresters before initiating trials.

Table 2-5 lists some of the general features of herbicide products that are useful in understanding uses and limitations. Table 2-6 lists the various general groups of plants on which herbicide use is common, and shows in a general way which herbicides may be possible choices. The charts shown in Figs. 2-2, 2-3 and 2-4 illustrate important species controlled by various herbicides in three regions of the United States. More

TABLE 2-5. Properties of Herbicides Used in Field Forestry (1979 Basis)

HERBICIDE (Common name)	FORMULATION	SEASON OF APPLICATION	CARRIER AND VOLUME	APPLICATION[b/] RATE
Amitrole-T	Amino triazole + ammonium thiocyanate liquid	May-July	10-15 gal/A in water	½ to 1 gal/A (1-2 lbs, a.i.)
Asulam	Asulox-sodium salt liquid	June-August after full frond development	10-20 gal/A in water	1 gal/A (3 lbs, a.i.)
Atrazine	80% wettable powder or 4#/gal flowable	Late winter	10 gal/A in water	3 to 4 lb ai/A
Dalapon	74% sodium and magnesium salts-water soluble	Late winter to early spring after grasses emerge	5 to 10 gal/A aerial, 10 to 100 gal/A ground in water	3 to 11 lb ai/A
Dicamba	Dimethylamine salt	All	Water or undiluted	Undiluted or 1:4 in water
Dicamba[c/]	Dimethylamine salts of dicamba & 2,4-D or 2,4,5-T	Spring to mid summer	15 to 300 gal/A in water	1 to 3 gal/A
	Oil-soluble acid of dicamba + isoactyl esters of 2,4-D or 2,4,5-T	Dormant (fall to late winter)	30 gal/A in diesel	1 gal/A
Dinoseb	Emulsifiable dinitrophenol	Spring to late summer	1 to 2 gal/A water or oil	1 to 2 gal/A (2⅔-5⅓ lbs, a.i.)

TARGET SPP. SELECTIVITY	RELATIVE PERSISTENCE[a/]	USE PRECAUTIONS	ROUTE OF UPTAKE	COST[b/]
Salmonberry and elderberry; will damage Douglas-firs if applied too early or late	Short-Med.		Foliage	$13/gal
Bracken fern; use on Christmas trees when not actively growing	Short	Apply when conifers are not actively growing;	Foliage	$31/gal
Annual grasses and some forbs; does not damage conifers when properly applied	Medium	Requires at least 2 inches of rain after application;	Root (+ slight foliage)	$2.50/lb
Annual and perennial grasses for site preparation; use with atrazine or directed for release	Short	Use ½ to 4 pints surfactant per 100 gal; delay planting 2 weeks if rate over 8 lb; apply when grasses are actively growing;	Foliage and root	$1.90/lb
Hardwoods and conifers by injection	Medium		Cut surface	$34/gal
Shrubs and weed trees for site preparation	Medium	Not as effective as picloram or as oil-soluble formulation	Foliage	$35/gal 4#/gal amine
Shrubs and weed trees for site preparation	Medium	More effective than water-soluble formulation	Stem	$17 to $20/gal
Nonselective, nontranslocated desiccant used to prepare herbaceous and woody vegetation for burning	Short	Must burn within 1 month of treatment; will not control resprouting; toxic; LD_{50} 58 mg/kg; absorbed through skin	Foliage	$9/gal

TABLE 2-5. Properties of Herbicides Used in Field Forestry (Continued)

HERBICIDE (Common name)	FORMULATION	SEASON OF APPLICATION	CARRIER AND VOLUME	APPLICATION[b/] RATE
Fosamine	Krenite-water soluble liquid	Late summer to early fall	10-40 gal/A aerial, 50-300 gal/A ground in water	¾ to 3 gal/A (3-12 lbs, a.i.) a.i.)
Glyphosate	Isopropyl amine salt	Herbs: spring Pre bud-burst Woody: fall, senescence	Water, 3-10 gal/A	¼-1 gal/A (0.75-3 lbs, a.e.)
Hexazinone	90% soluble powder; 2#/gal liquid Gridballs, 90%	Spring	Water 10 gal/lb per acre, 10 gal/A, None	1-3 lbs, a.i./A ½-1½ gal/A 4-8 foot spacing
MSMA	Monosodium acid methanearsonate - water soluble	Fall and winter	None	Undiluted (6 lbs, a.i./gal)
Picloram	Potassium salt + invert emulsions of 2,4,-D or 2,4,5-T[c/]	Spring to mid-summer	15-25 gal/A invert emulsion	1 to 4 quarts picloram + 1 to 4 gal of phenoxy invert
	Triisopropanolamine salts of picloram & 2,4-D (Tordon 101R & Tordon 101)	All	None	Undiluted (Tordon 101R is about the same as half-strength Tordon 101)
	Triisopropanolamine salts of picloram & 2,4-D (Tordon 101) with or without low volatile esters of 2,4,5-T or silvex	Spring to mid-summer	10 to 25 gal/A in water	1 to 4 gal/A
	Isooctyl ester of picloram + PGBE ester of 2,4.5-T (Tordon 155)[c/]	Dormant to budbreak	20 to 40 gal/A in diesel	½ to 1 gal/A

TARGET SPP. SELECTIVITY	RELATIVE PERSIS-TENCE[a/]	USE PRECAUTIONS	ROUTE OF UPTAKE	COST[b/]
Deciduous species for site preparation	Short	Applied in 2 month period before fall leaf coloration;	Foliage	$30/gal
Herbs, deciduous shrubs & hardwoods. Some ferns, moderately selective for conifers	Short	Do not disturb vegetation before treating.	Foliage	$52-$60/gal
Herbs, brush	Med.-Long	Low toxicity - Use low rate in sandy soil	Foliage & roots Roots	$16/lb ai
Hardwoods and conifers, by injection	Short	Wear protective clothing;	Cut surface	$14.50/gal
Shrubs and weed trees for site preparation	Long	Delay planting 8 months; must be used with Dow invert emulsions; RESTRICTED USE	Foliage	$63/gal + invert
Hardwoods and conifers by injection	Medium	May damage untreated conifers ("flashback"); Tordon 101R contains half the picloram of Tordon 101 NOT RESTRICTED	Cut surface	$11.50/gal for Tordon 101R & $21.48/gal for Tordon 101
Shrubs and weed trees for site preparation	Long	Must use application methods that reduce drift-RESTRICTED USE PESTICIDE	Foliage	$21.40/gal
Shrubs and weed trees for site preparation	Long	For use in California, Oregon, and Washington only; delay planting 3 to 6 months RESTRICTED USE PESTICIDE	Stem	$49/gal

TABLE 2-5. Properties of Herbicides Used in Field Forestry (Continued)

HERBICIDE (Common name)	FORMULATION	SEASON OF APPLICATION	CARRIER AND VOLUME	APPLICATION[b] RATE
Pronamide	50% wettable powder	October to December	10 gal/A in water	1 to 2 lb, a.i./A
Silvex[c]	Low-volatile esters (BEE, PGBE)	Late winter to summer	10 gal/A in diesel, water, or oil-in-water emulsion	¼ to ¾ gal/A (1-3 lbs, a.e.)
Simazine	80% wettable powder	Fall	10 gal/A in water	3 to 4 lb a.i./A
Triclopyr	Amine, 3#/gal; emul. ester, 4#/gal	May-August	10 gal/A water Ester may be mixed in oil or water	⅔ - 2 gal/A (2-6 lbs, a.e.)
2,4-D	Amine (for injection)	Spring and summer	None	Undiluted or 1:1 with water
	Low volatile esters (Iso-octyl, BE, PGBE)	Late winter to summer	5 to 20 gal/A in diesel, water, or oil-in-water emulsion	¼ to ¾ gal/A (1-3 lbs, a.e.)
2,4,5-T[c]	Low-volatile esters (Iso-octyl, BOE, PGBE)	Late winter to late summer	5 to 20 gal/A in diesel, water, or oil-in-water emulsions	¼ to ¾ gal/A (1-3 lbs, a.e.)
	Amine	Spring to late summer	None	Undiluted or 1:1 with water

[a]Short = ½ life $<$ 1 month; medium = ½ life, 1-6 months; long = ½ life $\leq$ 6 months, under humid temperate conditions

[b]ae = acid equivalent; ai = active ingredient; costs 1979 basis, 5-gallon quantities

TARGET SPP. SELECTIVITY	RELATIVE PER-SISTENCE[a/]	USE PRECAUTIONS	ROUTE OF UPTAKE	COST[b/]
Grasses only for site preparation or release in Christmas trees	Medium	Christmas trees ONLY. State labels. Check local status.	Root	$13/lb
Shrubs, weed trees, and forbs; slightly more damaging to conifers than 2,4-D or 2,4,5-T	Medium	Silvex is not a direct substitute for 2,4,5-T	Stem and foliage	$18/gal at 4 lb ae/gal
Annual grasses & some forbs for site preparation and release in Christmas trees	Long	Actively similar to atrazine; requires rainfall to activate	Root	$3.50/lb
Effective on maples, evergreen brush, some conifer tolerance	Short-Med.	Some soil activity	Foliage Stems Cut surface	$45/gal for 3#/gal amine
Hardwoods except cherry and big-leaf maple by injection	Short		Cut surface	$6.50/g at 4 lb ae/gal
Shrubs, weed trees, and forbs for site preparation and conifer release (except pines)	Short	May be used in combination with 2,4,5-T as brush-killer	Stem and foliage	$8/gal at 4 lb ae/gal
Shrubs, weed trees, and forbs for site preparation and release	Short	Can be used to release pines in late summer; also available combined with 2,4-D as a brush-killer mix.	Stem and foliage	$17/gal
Hardwoods by injection	Short		Cut surface	$19/gal

[c/]Registration status uncertain at time of printing. Products suspended 3/1/79, pending cancellation proceedings. Technical evidence appears to support continued registration for forestry uses.

Table 2-6. Herbicide Selection Guide

VEGETATION TYPE		PURPOSE
HERBACEOUS	Grasses and forbs	Site preparation
		Conifer release
	Ferns	Site preparation
		Conifer release
SHRUB	Evergreen	Site preparation
		Conifer release
	Deciduous shrubs	Site preparation
		Conifer release

AREA AND USE RESTRICTIONS	METHOD OF APPLICATION	HERBICIDE
Note: For all states with Forest Practices Acts, see rules of individual states		
Delay planting 2 weeks U.S.	Aerial or ground	Dalapon
None	Aerial or ground	Atrazine
Broadleaf forbs only	Aerial or ground	2,4-D[1]
Oregon, Washington	Aerial or ground	Hexazinone
Maine, Oregon, Vermont, Washington	Aerial or ground	Glyphosate
Oregon and Washington	Aerial or ground	Hexazinone, Pronamide
All U.S.	Aerial or ground	Atrazine Simazine
Bracken fern	Aerial or ground	Glyphosate, asulam
Sword fern	Ground	Dicamba
Christmas trees; bracken fern and reforestation	Aerial or ground	Asulam
Reforestation, bracken only	Aerial or ground	Glyphosate
	Aerial or ground	2,4-D 2,4,5-T[1,2] Silvex[1] Triclopyr ester
Delay planting six months	Aerial or ground	Picloram[2] + 2,4-D Dicamba + 2,4-D
See labels for seasonal use patterns	Aerial or ground	2,4-D 2,4,5-T[1,2] Silvex[1]
Preburn desiccation; Washington and Oregon	Aerial	Dinoseb
General brush control, all U.S.	Aerial or ground	2,4-D 2,4,5-T[1,2] Silvex[1] Picloram[2] + 2,4-D Dicamba + 2,4-D Glyphosate
Oregon, Washington		Amitrole-T, Fosamine
Use only 2,4,5-T to release southern pines	Aerial or ground	2,4-D 2,4,5-T[1,2] Silvex Triclopyr

TABLE 2-6. Herbicide Selection Guide (continued)

VEGETATION TYPE		PURPOSE
SHRUB (continued)		Conifer release cont.
WEED TREE	Hardwood	Site preparation

complete charts may be made up for local conditions and data. Local seasonal patterns are important (Fig. 2-5)!

The various properties of herbicides permit their use in several important silvicultural operations. Site preparation, release and stand improvement are all operations where herbicides can be used effectively.

Site Preparation

Site preparation entails the control of vegetation and formation of suitable microsites for establishing desirable seedlings. Non-selective herbicides may be applied without concern for desirable species unless there are enough desirable specimens in the area to be worth saving. In these cases, selective herbicides may be used to retain advanced regeneration or establishment trees.

Herbicides may be used in site preparation for control of grasses, broadleaf forbs, ferns, shrubs and undesirable hardwood competitors. The chemicals should generally be applied for this use when they have their greatest activity on target species. That is, herbicides should be scheduled so that they are most

AREA AND USE RESTRICTIONS	METHOD OF APPLICATION	HERBICIDE
Washington and Oregon: salmonberry and elderberry; Douglas-fir only	Aerial or ground	Amitrole-T
Washington and Oregon		Fosamine
All U.S.		Glyphosate
All U.S.	Aerial or ground	2,4-D 2,4,5-T[1,2] Silvex[1] Triclopyr
Delay planting 6 months	Aerial or ground	Picloram[2]
Deciduous species only (Pacific Northwest)	Aerial or ground	Fosamine
Deciduous species		Glyphosate
All U.S.; only 2,4,5-T to release Southern and Western pines	Aerial or ground	2,4-D 2,4,5-T[1,2] Fosamine Glyphosate Silvex[1] Triclopyr

effective rather than most selective, which is normally late spring to early midsummer for phenoxys, late fall for glyphosate, etc. Soil residual herbicides such as the triazines should always be applied to give the most complete weed control; because they do not usually affect conifers, whether applied before or after planting. Only in the case of picloram is it necessary to wait a substantial interval between spraying and planting. Restrictions of this kind will appear on the label.

The use of herbicides for site preparation produces certain results unique to herbicide use. Some of these are disadvantages, and it is important to minimize them.

Herbicides seldom, if ever, remove all vegetation in forest communities. Removal of only part of an undesirable community will result in the formation of a resistant community that may need re-treatment later. It is always desirable to avoid this. The species resistant to selective treatments are usually difficult to control later with treatments that are safe on crop species. Herbicide treatments should be selected for site preparation so as to minimize the emergence of species resistant to selective chemicals.

Figure 2-2. Susceptibility of northern forest species to aerially applied herbicides.

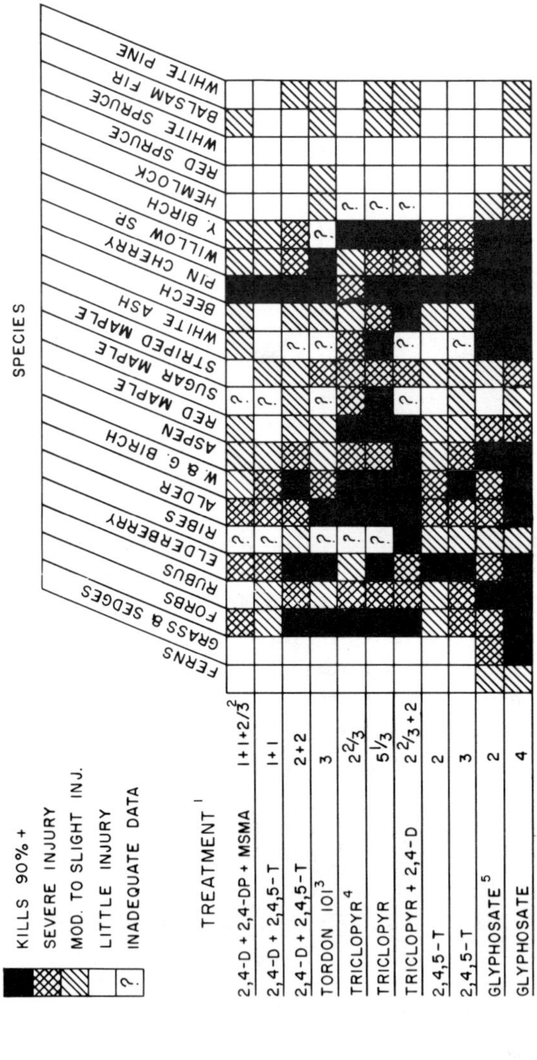

1 ALL TREATMENTS APPLIED MID-LATE AUGUST
2 NUMERALS REFER TO QUARTS PER ACRE OF INDUSTRIAL GRADE PRODUCT
3 A MIXTURE OF 2 LBS. 2,4-D AND 0.54 LBS. PICLORAM PER GALLON
4 GARLON, THE AMINE SALT OF TRICLOPYR IN A 3-POUND-PER-GALLON PRODUCT
5 ROUNDUP®, CONTAINING 3 POUNDS PER GALLON OF GLYPHOSATE ACID

Development of resistant communities may result from poor control of minor but resistant woody plants, or genetic selection in herbs. There is no expectation of genetic selection in woody species because of their long reproductive cycles and eventual suppression by crop species. Selection in herbs can only occur with repeated treatments. There is no evidence in any species of the creation of resistant or other "super" strains as the result of mutation or other herbicide action. Development of resistance is the result of repeated removal of sensitive types coupled with reproduction by those only slightly affected.

Another disadvantage of herbicides when used alone is the residue of standing dead or dying vegetation. This may be an obstacle to tree planters, and may also provide cover for elevated numbers of browsing or other damaging mammals. These problems can be overcome by combining fire or various mechanical procedures to reduce this residue to tolerable levels. The additional procedures will have additional costs, however, and the additional cost must be balanced against the reduced cost of planting and improved plantation growth. If the net cost per established tree is increased by the additional practice, it is not justified economically.

The use of herbicides alone has certain advantages for plantation site preparation. The total absence of physical soil disturbance substantially reduces the germination and development of pioneer vegetation. Avoidance of any procedure that breaks the duff layer will substantially reduce the onset of surprise weed problems. It will also prevent overstocking that may result from seeding by nearby desirable species. Proper site preparation favors the seedlings that are being planted, and restricts development both of weed species and of superfluous numbers of the crop species.

Total herbaceous weed control is desirable for plantations in areas likely to encounter severe summer drought. There are occasions where total weed control is also necessary for rodent control. The effect of herbicides in producing a clean weeded area cannot be reproduced with non-chemical means. Hand or machine scalping creates a temporary opening, but doesn't have residual effects. Furthermore, it often aggravates the weed problem by providing a seedbed for weed seeds, leading to heavy cover and rodent damage. Season-long weed control resulting from spring-applied residual herbicides is much more beneficial than the localized and more temporary effects of hand or machine scalping.

Figure 2-3. Susceptibility of southern forest species to aerially applied herbicides.

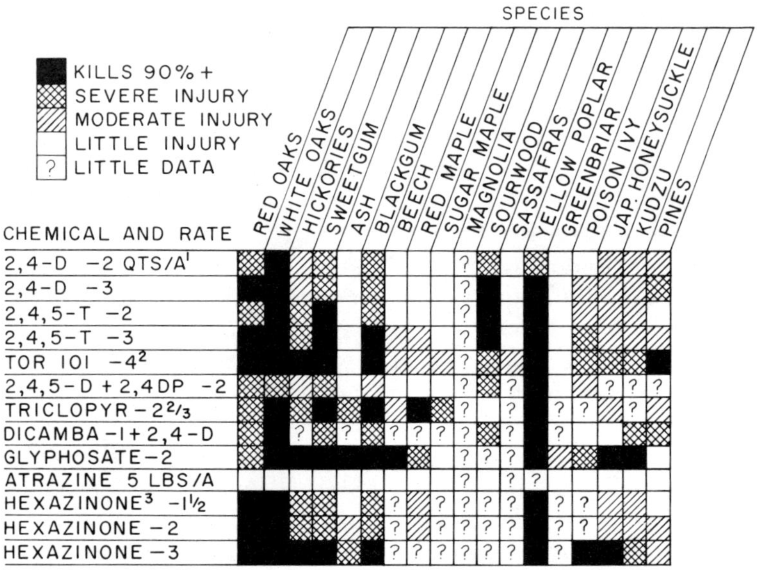

[1] RATES IN QUARTS PER ACRE LIQUID PRODUCTS, LBS ACTIVE PER ACRE DRY PRODUCTS

[2] TORDON 101 CONTAINS 0.54 LBS PICLORAM AND 2 LBS 2,4-D/GAL.

[3] HEXAZINONE FORMULATED AS GRIDBALLS, APPLIED DRY

Release

Release is the selective control of competing forest weeds in a stand of crop trees. Release may be achieved by any of several methods of applying herbicides, or by hand cutting, and includes both "cleaning" and "liberation" operations.

Several factors are important when using release-herbicides to maximum advantage. Of these, the most important is timely application. Release should be done early in plantation development, before there is significant shading of desirable trees. After shade has become dense, etiolation and root starvation of planted trees becomes progressively more severe. This leads to delayed height growth response while the released stand develops both roots and stem caliper. Saplings that have sustained prolong suppression have limited ability to withstand environmental exposure. Current evidence suggests that maximum growth of a tree is achieved if it is dominant through its entire lifetime. For most plantations established in clearcuts, release during the second or third year after clearcutting and planting will prevent most suppression problems, and will require the least herbicide and cost per unit of benefit. Delays lead to loss of growth, sunscald lesions and disease, and breakage. Good site preparation reduces needs for release, and a good release program reduces the need for intensive site preparation. A good combination of site preparation and release treatments usually minimizes overall disturbance.

Scheduling of release treatments is extremely important. Release operations capitalize on basic biochemical selectivity among species. All herbicides can injure crop species, but injury can be kept to a minimum by selecting the season for maximum selectivity rather than the season of maximum effect on target species. Sometimes the season is the same. For phenoxy herbicides, the season for release in the western states is either early spring or after terminal growth is completed, the so-called dormant seasons. In the South, phenoxys must be applied while pines are in active growth in order to control target species. For glyphosate, selective treatments are before growth begins or in late fall in the North, depending on whether herbs or woody species are to be controlled, but mid-summer applications may be safe in the South. Each herbicide has its own set of characteristics which determine season of application. Patterns are usually similar for the herbicides within chemical classes, but there are exceptions. For example, simazine is normally applied in fall, whereas the more soluble atrazine is applied in spring, except in dry areas. The reason for the difference (solubility) will not necessarily appear on the label, so it is always important to follow specific label instructions.

Different kinds of vegetation and different seasons of application may require different solvents, or “carriers” if phenoxy herbicides are used. Spring dormant applications in the Pacific Northwest require a diesel oil carrier if they are to be applied to deciduous species. If they are to be applied to evergreen species, an emulsion of water and oil may be suitable, or water alone for some species. Oil should be added to the water carrier for fall treatments with phenoxy herbicides, as well. Oil will be prescribed if a substantial proportion of the herbicide must be absorbed through bark in ground or aerial applications. This is not a consideration with soil-applied herbicides, or with those that are completely soluble in water but not oil. They are applied in water at any time. The use of oil may be foreclosed eventually by scarcity. In this event, the spring dormant treat-

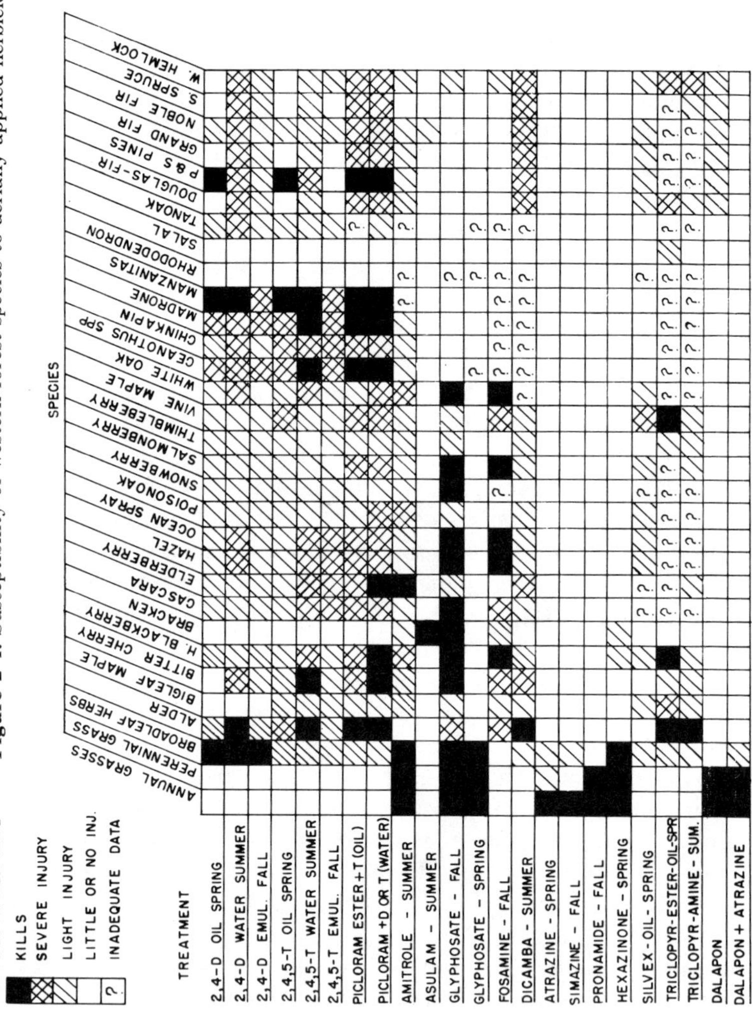

Figure 2-4. Susceptibility of western forest species to aerially applied herbicides.

ments will lose effectiveness, because there are no known substitutes with similar properties.

Herbicides have certain advantages over hand release. Herbicides provide relatively slow release that persists for many years. If the stand has developed too far under shading vegetation, it will have shade-type foliage and thin bark. Both shade-leaves and thin bark are sensitive to direct sunshine, and sudden release by hand clearing will cause sunscald and defoliation of old leaves, leading to delayed growth response. The residual shade from standing dead stems and foliage provides a slow transition from shade to sun, and provides the released tree an opportunity to develop sun-leaves and to cover its thin bark with foliage. Growth response appears quickly. Sunscald and defoliation are rare after herbicide treatment.

Another advantage of chemical release is the lack of physical disturbance. There are no changes in the environment of a chemically released tree other than increased quantities of radiation and soil resources, both are desirable. Totally absent are the impacts of falling debris and incidental saw damage characteristic of hand release. Debris impacts are of little importance when stands are released in their very early years, but are serious problems after stands have reached the age of 5 or more years. In stands 10 or more years old, debris fall is serious after herbicide release, as well, so release should not be delayed.

Commercial timber species vary substantially in their need for release. When anticipating the appearance of a weed problem requiring release at the time of planting, it is wise to consider planting a species that does not need such complete freedom from competition. For example, in the Pacific Northwest, western hemlock planted in place of, or in mixture with, Douglas-fir will give some assurance that a stand will develop even if it doesn't receive a release treatment. Should the need for release appear, a lighter treatment will suffice for shade-tolerant hemlock. Red or black spruces and balsam fir play such a role in the Northeast. Shortleaf pine is suitable in parts of the South, and true firs and Douglas-fir function effectively in this role elsewhere in the West.

There are times when a small amount of damage to the crop species results from a chemical release treatment. The occurrence of light damage will not harm the trees permanently. Herbicide injury is transient in most conifers when it does not exceed partial defoliation, and growth loss will be small. Even the loss of occasional growing tips or leaders does not signify a major loss of growth. If the leader is injured and the laterals are intact, the loss of a partial year of growth and occasional forked trees are the only adverse consequences of injury. These impacts are more than offset by the benefits of controlling the competing vegetation, especially when the alternatives may lead to loss of the entire stand.

Timber Stand Improvement

Improvement in the composition and stocking of older stands has traditionally been accomplished with axes and saws. The advent of selective herbicides applied by helicopter, and development of water-soluble formulations and equipment for tree injection, have relegated the saw to a minor role.

Injection is usually selective when herbicides are used which lack soil residue activity. When using herbicides such as picloram, that have residual properties, proper dosage and careful handling can minimize damage to crop trees. Pines and some hardwoods may be affected by root transfer, or flashback, to desirable untreated trees. The appropriate selection of herbicide, dosage and timing of application can provide complete control of all stems treated with virtually no unanticipated adverse effects. This leads to the development of a nearly ideal stand in one operation if adequate stocking of desirable species is present initially.

Aerial application of selective herbicides has some potential uses for stand improvement. For example, the phenoxy herbicides are not only selective between hardwoods and conifers, but also among hardwoods. In the Northeast, it is possible to increase the proportion of sugar maple and ash species in stands of poplar, birches and red maple by applying low rates of 2,4,5-T in August. Aerial application of herbicides should be regarded as an interim measure, however, and serve only to release desirable species pending more precise selection and spacing with injection or other methods.

Application Methods

Herbicides may be applied in forests by aircraft, backpack sprayer, backpack mist blower, tractor-mounted mist blower, tractor-mounted boom sprayer or tree injectors. Each method has certain attributes when using herbicides that do not extend to insecticides, so a separate discussion of methods is warranted here.

Foliage Sprays.

Most herbicides are applied to foliage. A key feature of foliage sprays is that the active ingredient be distributed as thoroughly

Figure 2-5. Injury to short-needled conifers, western pines and major brush species in relation to timing of application. Pacific Northwest. 2,4,5-T, two pounds acid equivalent per acre.

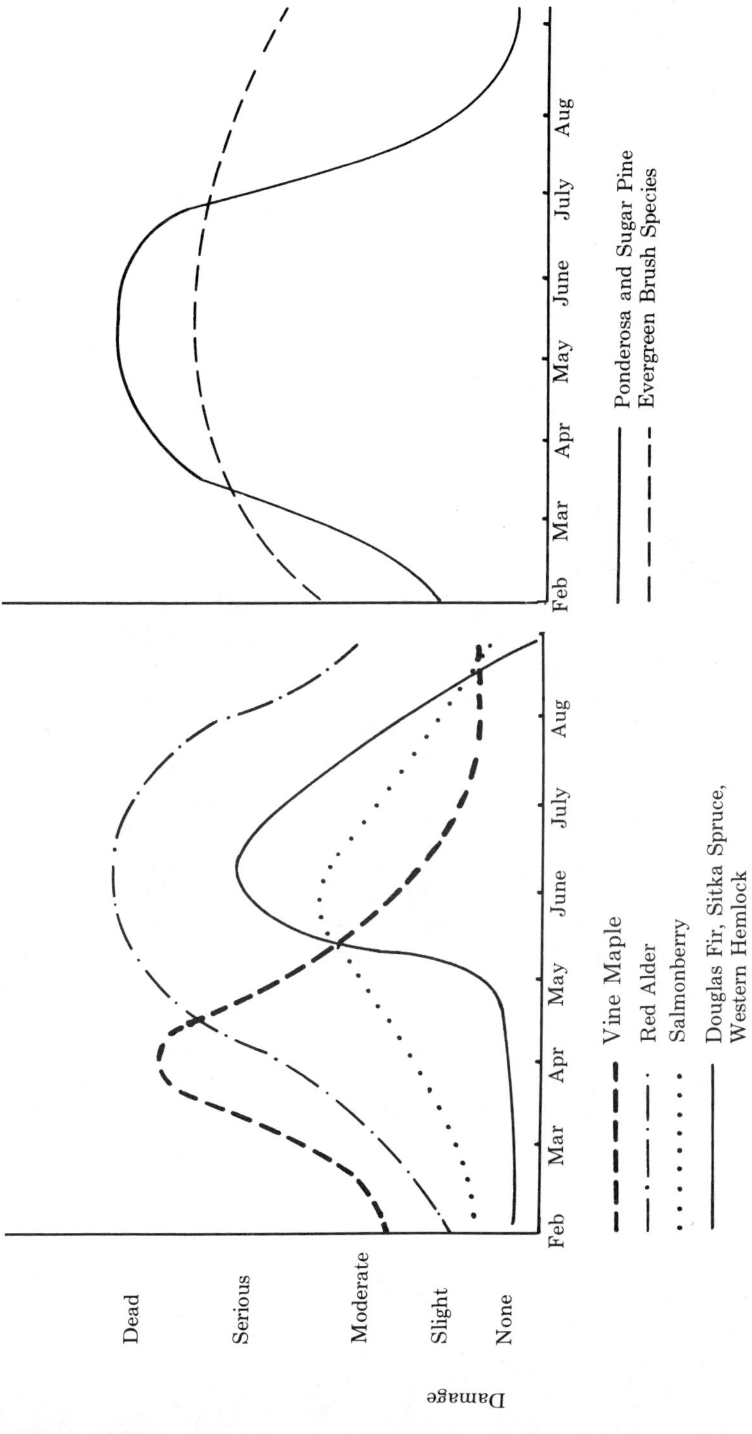

as possible on the receptor leaves without causing heavy deposits on the ground or losing large amounts to atmospheric drift. Forest foliage typically forms the equivalent of several thicknesses of leaves between the top of the canopy and the ground. This creates coverage problems not encountered in agricultural crops.

An ideal spray for best coverage would have an infinite number of droplets of infinitely small size that would lazily drift around among the leaves until all the active ingredient was in contact with leaves or stems. This kind of spray obviously cannot be developed; if it were, off-target drift would be an insurmountable problem. So all sprays are compromises between the ideal type of coverage, and the ideal type for drift control, which would have no fine droplets, and would cover poorly.

Drift occurs when air movement carries fine droplets away from the target area. The amount of drift is determined by the length of time the drop is in the air, which is inversely (but not precisely) related to the square of its diameter. In winds of less than five miles per hour, drops with a diameter of more than one-fifth of a millimeter (200 microns) will land within fifty feet of the point of release if applied within twenty feet above the target. Smaller droplets will be carried further by the wind because they take longer to settle. Nozzle systems are therefore designed to keep as much as possible of the spray volume in drops more than 200 microns in diameter for maximum safety and effectiveness in foliage spraying with herbicides.

Large droplets require high volumes per acre to achieve the same number of droplets per square inch. There is, therefore, much incentive to have droplets as uniform as possible, and as close to 250 microns as possible. As a general compromise, volumes of 4-10 gallons per acre are applied with conventional nozzles for nearly all forest herbicide applications by aircraft. When coupled with careful application procedures and strict weather criteria, these procedures are adequate to virtually eliminate drift. Volumes below five gallons per acre are suitable only for wettable powders and highly translocated materials. Tractor-mounted ground rigs normally apply at 20 or more gallons per acre when using a boom. Mist blowers usually apply five to eight gallons per acre. When using a hydraulic spray gun, on foliage, sprays are generally applied until foliage is thoroughly wet. For further details on application technology, see Chapter 4.

Flexibility of methods varies with season. Sprays must be applied to foliage in several seasons of the year. Early foliar sprays are timed so that the first leaves to appear are expanded at the time of application. New leaves are still emerging at this time and conifer buds have normally burst. Timing for early foliar treatments is critical. Postponement until conifer buds are open more than five centimeters (2 inches) will result in excessive damage to new foliage; treatment before the first hardwood leaves are fully expanded will result in very poor brush control. For this reason, treatments during the early foliar sprays have high risk of not being finished due to bad weather. The advantage of applying in this season is that it controls certain species needing spring treatment without using oil.

Summer foliage treatments are normally used for site preparation. The early part of this season is also used for releasing southern pine. This is the season of greatest effect on brush and hardwood for picloram and phenoxy herbicides, and it is also the season of greatest damage to conifers other than southern pines. The summer foliage season extends from the time foliage is fully expanded until the termination of conifer leader growth, roughly late July. Summer foliage sprays with phenoxys, picloram and amitrole have the advantage of providing for minimum re-sprouting. It is also a season when desiccation may be accomplished satisfactorily in preparation for burning. It has the decided disadvantage of offering maximum drift opportunity, both in terms of sensitivity of crops and elevated temperature. Where drift-sensitive crops are nearby, 2,4-D should not be used and drift control procedures should be considered.

Late foliar spraying extends from early August to two weeks before leaf fall; it extends later in evergreen hardwood and brush types. The late foliar season is used for pine release with 2,4,5-T in the Northwest, and for release of pine and spruce in the Northeast. Glyphosate and fosamine are applied for woody plant control exclusively in the second half of the late foliar season, at which time they are used for both site preparation and release. The phenoxy herbicides are not as effective during the late season as they are earlier. This timing is critical to provide selectivity for western pines, however, which are normally quite sensitive during the early dormant season.

Dormant Sprays.

Spring and fall dormant spraying are generally used exclusively for conifer release. Spring dormant treatments are limited to the Pacific Northwest, where certain deciduous brush types are highly sensitive to 2,4,5-T in oil, and where evergreen brush types are effectively controlled by an emulsion of water and oil. This timing is especially well suited to Douglas-fir, western hemlock and Sitka spruce release but is extremely damaging for the western pines (Fig. 2-5). As a general matter, the spring dormant treatment provides only topkill and stimulates sprouting much as cutting would. This season is therefore unsuitable for site preparation unless picloram ester is mixed with 2,4,5-T. This combination will destroy the conifers, but will inhibit sprouting of the hardwoods as well.

Fall dormant treatment is simply an extension of the late summer foliar season. Its use is limited to pine release in western evergreen shrub and hardwood types.

The soil residual herbicides, atrazine, simazine, dalapon, pronamide, and hexazinone, are all used during dormant seasons. Simazine and pronamide are normally applied in fall, and the others are normally applied in spring. There are local exceptions, however, where timing of precipitation is the controlling factor. All these herbicides need adequate rainfall to move them into soil where they are taken up by roots. They may be degraded or adsorbed (bound by chemical attraction) on organic matter or soil particles if applied during the dry season, or if adequate rainfall fails to materialize between application and the time weed control is expected.

Basal and Stem Spraying.

Backpack sprayers filled with herbicide-in-oil mixtures are used for controlling hardwoods under a variety of circumstances. Basal treatment is the application of an oil solution to the basal portions of the stem. Chemical is applied to the point of runoff. It is important to cover the entire circumference with a saturating coverage on the bottom 15-30 cm (6-12 inches) of the trunk. The mixtures used in this treatment are higher in concentration than those used for normal foliar treatment with the same equipment, and the method provides a direct targeting of nearly all material. For these reasons, the treatment is highly effective and virtually eliminates sprouting. Silvex and 2,4,5-T are very effective when applied as basal sprays; 2,4-D + 2,4,5-T brush killer mixtures are also used.

Basal and stem treatment utilize the ability of the oil carrier to penetrate bark, thereby facilitating entry of the herbicide into the vascular system of the tree. Foliage is not necessary for success, and the method may be used at any time. Greatest effectiveness, however, is observed in fall treatments when movement of the phloem contents is generally toward roots.

The basal or stem treatment is costly in terms of oil and herbicide. It is also labor intensive and leads to maximum (although apparently quite safe) worker exposure to herbicide in a difficult-to-remove oil. The excessive costs have caused this method to be replaced by stem injection for most silvicultural operations.

Stem Injection.

Stem injection entails the use of hatchet-type or tubular tree injector to create a wound in the inner bark of the tree into which a herbicide mixture is placed. Through the years injection has evolved through phases in which sodium arsenite was placed in axe cuts, oil solutions were placed in frills, and now water-soluble systemic herbicides are placed in small wounds at some spacing apart at any convenient height. All these treatments are effective, but convenience, safety and cost have led to steady development of new technology.

Herbicides used for stem injection should be soluble in water. Derivatives of 2,4-D, 2,4,5-T triclopyr and picloram suitable for injection are the amine salts. These materials and the sodium salts of cacodylic acid and MSMA are normally used undiluted at the rate of roughly one milliliter (¼ teaspoon) per injection. Other water-soluble herbicides may eventually be registered for stem injection, and several are known to be effective for this work. Sodium arsenite is no longer registered because of toxic hazard to applicators and wildlife.

Dosage used in stem injection is usually controlled by spacing of cuts. Small trees and sensitive species may be controlled with injections 15 cm (six inches) between centers. Larger trees and resistant species may require a spacing of 7½ cm (three inches) between centers (Fig. 2-6). Very large cull hardwoods may require complete frilling for effective control. Automatic equipment cannot handle fluids of high viscosity. Chemical may be diluted somewhat to reduce viscosity in winter or for treatment of sensitive species. Use water only for dilution. The concentration in the salt solution will prevent freezing at dilutions adequate for reducing viscosity. Do not dilute further.

Tree injection may be done at any time of year. It is important to use the correct herbicide for the current season, however, and a year-round program may require several different herbicides. In the northern hardwoods, for example, organic arsenicals may be most effective for fall applications, whereas

Figure 2-6. Distance between injections is an important factor in tree injection dosage. Large trees must be treated with injections closer together than small trees in order to achieve comparable effects. Treat below line crown.

picloram/2,4-D mixtures may be needed for early summer and winter applications. Injection is not usually recommended during the spring period of rapid sapflow. In the deep South, organic arsenicals are not adequately effective at any season, but 2,4-D amine is effective in spring and early summer except on maples. Treatments in the Appalachian area follow northern rules. Picloram and 2,4-D mixtures are usually used when 2,4-D or organic arsenicals are not effective. Triclopyr is also very effective.

Stem injection may be used for control of residual cull hardwoods, for stand improvement, or for pre-commercial thinning. It has advantages for each use. Injection of culls is an inexpensive method that is highly specific for the target trees and is low in labor costs. Removal of relatively small numbers of culls on large areas can be accomplished with little chemical and labor commitment; large numbers of small stems may be controlled at reasonable labor and chemical costs. The method is ideally suited for heterogeneous hardwood forests where many species may be present, and where aerial treatment would be unsatisfactory.

Stem injection is a useful approach for pre-commercial thinning of conifers in stands where average diameter is above 7.5 cm (three inches). The use of MSMA in fall or early winter will provide systemic killing of treated trees with negligible flashback. It is important to treat below the bottom of the live crown if total kill is desired.

Chemical thinning, as outlined above, has some biological advantages over cutting. In particular, it leaves the stand resistant to both wind and snow damage. It also protects against sunscald and defoliation from solar damage to shade leaves. The organic arsenicals, in particular, have the property of both killing the tree and of controlling both insects and root rots of some conifers. These effects are apparently produced by metabolites of the herbicide produced during degradation by microorganisms. The remarkable coincidence of degradation of the herbicide and control of most of the biological and physical problems of precommercially thinned stands, makes organic arsenicals and trunk injection an ecologically desirable treatment.

Stem injection provides some worker exposure to herbicides. The materials are water soluble and concentrated. They therefore have some opportunity for absorption by skin. Despite their low toxicity, crews should be trained to handle them with respect, and to follow instructions on the label.

Stump Treatment.

Stumps may be treated after tree felling either with basal sprays or with the herbicides used in injection in the same amounts. It is always best to treat stumps within five minutes after felling. Within two hours xylem vessels appear to become closed to absorption.

When using the water-soluble injection herbicides, the herbicide should be applied in a thin line around the perimeter of the stump just inside the cambium. No more than 8 milliliters (two teaspoons) of picloram-plus-2,4-D should be needed to severely inhibit sprouting from a 12-inch (30 cm) stump if applied correctly. Stumps may also be basal sprayed. Treatment during the period of heavy sap flow in spring may produce inferior results. Do not use organic arsenicals for stump treatment, because they do not translocate well into the roots after cutting.

A pelletized form of picloram has been developed for manual application around bases of cut stumps. Timing of application should be spring when rains will move the herbicide into the root zone at the time of maximum foliage development. This type of herbicide preparation may be the most practical method of dealing with stumps that have already sprouted. Users should be cautioned that conifers or other desirable trees with roots in the treatment zone may be injured or killed.

Other Methods.

Herbicides may be applied in various ways other than the conventional methods listed. Pelletized or granular herbicides, such as picloram or hexazinone, may be applied in some spacing pattern (e.g., a grid) to be taken up through weed tree roots a safe distance from crop tree roots. Organic arsenicals have been injected into bore holes. Sodium arsenite, no longer registered for use, has been used in strips of blotting paper inserted under bark, and so on. In all cases, it is important to follow directions for application on the current label.

Uses of Herbicides for Wildlife Habitat

Forest wildlife are dependent on the habitat provided by both dominant and subordinate species of vegetation. These have traditionally been subject to silvicultural practices which have fortuitously been beneficial for many species. There is mounting evidence, however, that wildlife habitat can be deliberately managed to benefit certain species, or to prevent certain species from becoming pests, while also meeting other land use objectives.

Diversity of habitat is generally accepted as leading to a diverse wildlife community. Many of the most popular game species are most abundant in "edge" communities where forest and clearing interface. They use the clearing for foraging, and reside in the forested area for cover. Patchwork clearcutting is well suited to these species; but poses economic disadvantages and some silvicultural problems if patches are of small enough size to be ideal habitat. Small clearcuts often give rise to competitive cover that is too extreme for regeneration or preferred timber species. This need not be a problem if the units can be sprayed with herbicides that remove the most serious competitiors while maintaining the browse and bud supplies. Phenoxy herbicides, applied early in the spring, will provide the needed release and will lead to prolific sprouting. This practice has been used extensively to promote browse production on winter range and is equally effective in promoting conifers.

Harvest and spray units within a given forest area should be staggered in time to provide a spectrum of age classes for wildlife diversity. This arrangement will insure a productive wildlife community, and provide for near-maximum yields of timber.

Herbicides may be used for maintenance of certain types of bird habitat. Harvesting or other clearing operations may lead to extensive areas with no residual trees for perches and hole nesters. Unmerchantable and low-grade hardwoods may be left standing without causing major silvicultural problems if they are injected or sprayed at the time of clearcutting. The herbicide treatment is normally preferable to girdling, because girdling may lead to too-vigorous sprouting, and the girdling ring may lead to weakening at the base and falling. Large cull trees, traditionally a problem for silviculturists, can thus be made a benefit for predatory and hole nesting birds. Retention of hollow trees in this way also provides habitat for racoons and other species. The cost of such treatment is normally lower than felling. The amount of herbicide used for this purpose is approximately a milliliter of chemical per 2.5 cm of tree diamter for large culls. Organic arsenicals are the preferred herbicides for this type of treatment because they do not favor rapid decay.

Areas managed for big-game winter range depend on good browse and adequate cover. Winter browse is typically most abundant in stands of sprouting species in the 3-15 year old age category after major disturbance. Winter browse areas do not remain productive indefinitely. Sprouts tend to grow out of reach, at which time the quality and quantity of browse decreases. Spring dormant sprays of phenoxy herbicides, primarily 2,4-D but also 2,4,5-T, are useful for rejuvenating shrub and hardwood communities grown out of reach. Normal spring dormant conifer release sprays are ideal for this. When coupled with widely spaced conifers, there is an extended period of highly productive browse. The use of triclopyr in summer will favor cedar browse because members of the *Cupressaceae* family appear to be resistant.

There are many other potential uses of herbicides for habitat improvement. The professional wildlife manager will have an understanding of the habitat requirements of species for which increases are desirable. By following the guidelines outlined in this section and in the tables given for silvicultural uses, a full range of prescriptions can become available for habitat management.

Habitat may be managed to prevent pest mammals from becoming epidemic. Forest plantations established in herbaeceous cover typically have problems with mice, voles and gophers. The small rodents girdle seedlings or feed on roots, especially under snow, and cause heavy losses. Present methods of reducing damage generally entail the use of highly toxic rodenticides. Habitat management should be more effective for reducing the basic cause of wildlife problems, and is also beneficial to trees because of the general effect of weed control. Numbers of animals increase again as habitat recovers.

The use of habitat manipulation to control an epidemic population can backfire. Animals deprived of a food supply will eventually leave or starve, but they can be expected to cause heavy damage to planted conifers before their numbers drop. A better strategy is to remove the preferred habitat in anticipation of planting by at least several months. This will reduce the numbers of pest mammals before the vulnerable seedlings are introduced. It may also be necessary to combine the herbicide program with a one-time baiting program.

The use of herbicides for mammal habitat management requires general habitat management. Spot treatment around planted seedlings for grass control will provide some protection from girdling, but will not affect total numbers of animals significantly. Spot treatment is also ineffective for pocket gophers; this animal will often burrow in the treated spots when soil is too dry elsewhere, with disastrous results for the conifer.

There is evidence that habitat management will influence animal populations as a general principle. Research with deer, rabbits, pocket gophers, voles, deer mice and certain shrews indicates that numbers of animals will decrease when habitat conditions change for the worse and increase when it changes for the better. Habitat management for pest control should, therefore, be regarded as a reversible and temporary measure to permit certain desirable species to escape heavy animal pressure without using animal toxicants.

Safety

Safety in the handling of herbicides is primarily a matter of avoiding damage to valuable plants. So much publicity has been given to environmental hazards and human health risk from herbicides, however, that these topics deserve consideration here. General information on pesticide safety is given in Chapter 5.

There is a natural tendency for people to mistrust chemicals. This is a natural response to our inability to see or feel what the chemical is doing. The herbicides used in forestry, however, have mostly been studied in minute detail, and the quality of information is good. They range in toxicity from "moderate" to non-toxic. They do not penetrate skin in harmful amounts even with extreme exposures, and most of the chemicals are eliminated rapidly. Two recent studies have determined that the most heavily exposed applicators receive between 1/200 and 1/400 of the maximum *safe* level. Additional evidence suggests that residents in sprayed zones are far less exposed.

If human contact with herbicides occurs, it is usually via dermal exposure. Oil-soluble materials do not move across skin membranes rapidly, but water-soluble amines and other salts may enter more readily. For this reason, it is appropriate to wear protective clothing in the form of long-sleeved shirts and gloves when handling the water-soluble concentrates. When using organic arsenicals, skin cream on exposed surfaces will minimize the unpleasant odor of absorbed herbicide that produces a garlicky sensation in the nose and mouth. A tank of clean water and change of clothes should be kept on hand where the concentrate is handled. Spills on skin should be minimized and washed as soon as possible after occurrence.

The alternatives that will achieve vegetation management goals without chemicals entail different risks, and at present do not enjoy a comparable safety record. Hand cutting, especially, is regarded as a hazardous industrial occupation. Bulldozer operations and other vegetation control options all have certain risks. Existing evidence suggests that risks of chemical and non-chemical methods are not strictly comparable, but that the net risk to humans from chemical methods is very low.

Of particular interest regarding environmental contamination is the likelihood of spills during loading of sprayers and of atmospheric drift. A spill on dry land is not likely to move very far beyond the apparent flow; the chemicals simply do not move readily in soil. If they should backsiphon into a stream, however, a downstream use of water for irrigation could cause damage to crops. It is, therefore, of the greatest importance to fill tanks with equipment that cannot allow the material in the tank return to the stream or pond.

The only known injuries to wildlife resulting from direct exposure of animals to herbicides has been from spills of organic arsenical herbicides. These materials are sodium salts, and are attractive as salt licks. It is imperative that injectors be filled carefully to prevent spills that might become salt licks. Should such a spill occur, it is important to have a shovel on hand to bury the contaminated soil under at least a foot of packed soil, or preferably, to haul the contaminated materials away for disposal at an approved disposal site.

Spills of large size are covered by laws of individual states in some instances. Herbicide users should become familiar with local laws. In the absence of such laws, the following suggestions are offered for spills of various sizes:

1) Spills of less than one pint (454 ml) of concentrate, pure or diluted but significantly more localized than intended on the label: Bury if focused on a spot, or dig up and remove to an approved disposal site. Place sign specifying herbicide, and when spilled, if diffused over less than 100 square feet and removal is not possible.
2) Spills of one pint (454 ml) to one gallon (3785 ml) of concentrate: Remove visible spilled material by shovel if focused. Haul to disposal site or spread in soil along ridge topography on own land where effects of herbicide will not harm desirable crops. Picloram should be confined whenever possible. If diffused so that removal is not feasible, dig shallow ditch around spill so that surface water will be directed into a hole at least two feet deep. Warning signs with chemical, amount and date of spill are suggested.
3) Spills larger than one gallon (3785 ml) of concentrate should be removed if possible with a shovel. If not, surround the spill with a ditch leading to a dry hole. It is important to prevent any surface runoff; movement through soil is limited. In addition, warn neighbors downstream that a spill has occurred and that leachate from the spill could affect water quality temporarily. Signs are suggested. Notify the local Agricultural authorities as to the nature and location of the spill and follow their instructions.

Atmospheric drift is a problem when several conditions occur *simultaneously.* Adjacent vegetation, or sensitive crops or ornamentals, must be nearby and downwind from the project. The applicator must release a high percentage of fine droplets $<$ 200 microns in his spray pattern, and atmospheric conditions must favor spray movement toward the sensitive area. Nearly all drift damage is caused by 2,4-D or picloram because minute quantities cause formative effects on plants that are readily visible.

Drift damage can be eliminated by avoiding the treatment of boundary areas when wind is moving toward the boundary. As a practical matter, aerial applications of herbicides with conventional equipment should be restricted to conditions of wind less than five miles per hour, and sensitive areas should be sprayed when conditions are cool, with a slight breeze away from the boundary.

Minimizing the Need for Forest Weed Control

We have given much attention to various ways of controlling forest weeds. This discussion would not be complete without a reminder that most weeds have become problems because of human activities of certain kinds. Weed problems are usually caused – they do not just happen. One obvious way to control weeds is to prevent their development in the first place.

The most widespread weed problems are the residual unmerchantable hardwoods left after harvesting. Removal of such material at the time of harvest is the best method of preventing this problem from developing. These trees can be felled, girdled or injected. This simple procedure is a fundamental part of constructive silviculture, and is applicable to even and uneven-aged management systems.

Another principle is that seral weed species thrive on disturbance. Any forest operation that causes excessive disturbance will be invaded by many pioneer species. Operations that remove timber without disturbing soil will bring about fewer weed problems. The vegetation manager is reminded that one does *not* want a good seedbed when planting precisely the stand that he wants to grow.

The forester has considerable control of the degree to which logging and other activities cause weed problems. The regeneration forester and the logging superintendent should be in close communication to insure that the logging operation does not lead to additional problems for regeneration. Good planning before harvesting will permit the harvesting and reforestation operations to work harmoniously in providing maximum productivity with the least cost and least undesirable side effects of either logging or reforestation practices.

The Vegetation Management Plan

The vegetation management plan is developed as a part of over-all silvicultural planning, including regeneration, young stand tending and harvesting. When vegetation is controlled in support of desirable regeneration, the vegetation management plan will usually be a part of the regeneration plan. If vegetation control is part of a wildlife management plan it should be part of that over-all plan not just an add-on.

A fundamental feature of forest land is that the *dominant trees control all other aspects of land use.* This principle holds for all land uses. Therefore, vegetation control that influences development of the dominant canopy therefore has a controlling influence on all other activities in that forest. Vegetation management is therefore the *primary* tool in implementing any management scheme.

The vegetation management interface with reforestation is important for many reasons. In particular, the choices of tree seedling size and species are both dependent on the type of habitat. Small trees will not do well in a jungle of standing dead material, but large trees will do very well. The vegetation management plan can save considerable disturbance and capital outlay if it substitutes herbicides for mechanical clearing or burning, but the reforestation specialist must have access to large planting stock of suitable species. Use of ill-adapted seedlings will cause the whole project to fail.

Vegetation management plans should be formulated well in advance. The ideal time for planning is before a mature stand has been harvested. Reforestation immediately after harvesting is a major component of vegetation control; if the planting is successful, release requirements will be minimal. Again, reforestation with large seedlings will reduce the release requirements even further. Failure to reforest immediately may set up a rehabilitation problem entailing higher costs and more treatments.

The plan itself should first specify land use objectives and describe the conditions prevailing at the time on the major management unit or working circle. Harvesting patterns, history of vegetation development and planned cutting activities are important preliminary information. These should be supplemented with lists of species affecting management objectives in an undesirable way and indicate at what time they are likely to be important. If serious brush and weed problems exist at the time, an inventory of desirable and undesirable species and their impacts should be developed. The background data for these inventories can be obtained with precision in a land examination or regeneration survey that records frequency and dominance of weeds as well as crop species.

The plan should encompass a substantial time interval (Fig.

***Figure 2-7.** Planning phases and capital outlay in vegetation management.

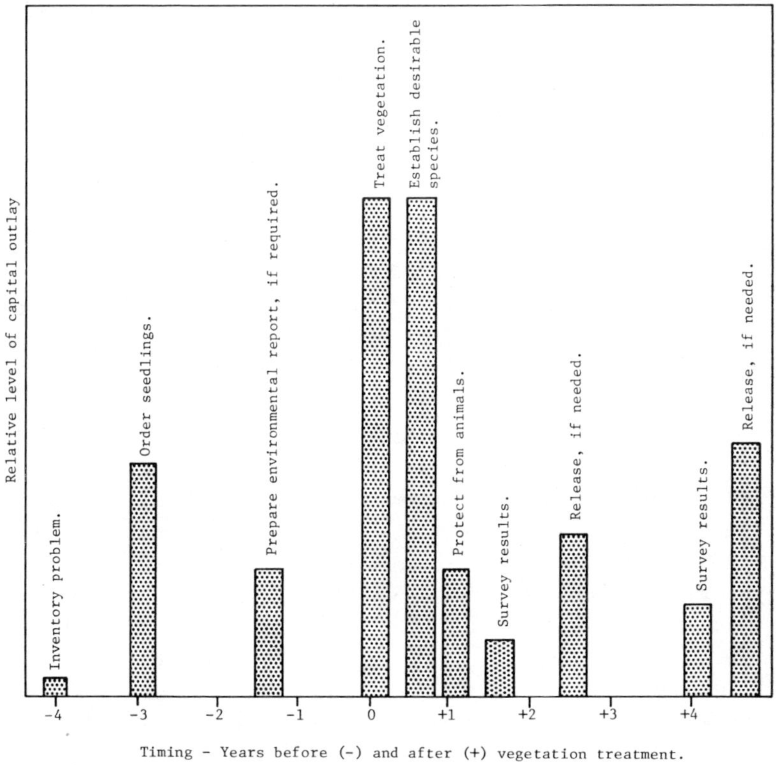

2-7). Vegetation management should be linked to reforestation efforts so that budgets for reforestation, site preparation and release can all be submitted as part of the same operation. Neither can succeed without the other, so the plan should be organized as an integrated program. The time sequence of operations should be formulated so that activities and resource needs can be anticipated several years in advance.

The vegetation management plan should identify sensitive areas, if any. There are several ways of treating most weed problems. If the sensitive areas are well defined, appropriate adjustment of methods will avoid trespass. Under some circumstances, alternatives of higher cost may be needed. Sensitive areas can be treated without off-site effects. Herbicides may be used near open water if applied without direct introduction into the water. But such applications may require specialized equipment or chemicals. An advance plan will insure that these areas do not get overlooked or deleted from the resource.

Chapter 2 References

Audus, L. J. 1976. Herbicides, physiology, biochemistry, ecology. 2nd edition, 2 volumes. Academic Press, New York.

Byrnes, W. R. and H. A. Holt, (Eds.). 1975. Herbicides in forestry. Department of Forestry and Natural Resources, Purdue University, Lafayette, Ind.

Crafts, A. S. 1961. Chemistry and mode of action of herbicides. Interscience, New York.

Gratkowski, H. 1975. Silvicultural use of herbicides in Pacific Northwest forests. U.S.D.A. Pacific Northwest Forest and Range Exp. Sta., Gen. Tech. Rep. PNW-37, 44p.

Lawrence, W. H. and J. D. Walstad. 1978. The ecological basis for vegetation management, p. 113-117. *In* USDA/USEPA Symposium on the use of herbicides in forestry. 213 p.

Newton, M. (Ed.). 1967. Vegetation management in forests, ranges and non-crop lands. Oregon State University School of Forestry, Corvallis, 346 p.

Newton, M. 1969. Herbicides and the tending of forests. *In* Woodland Handbook for the Pacific Northwest. C. R. Ross, Ed., Oregon State University, Coop. Extension Service, Corvallis.

Newton, M. 1975. Constructive use of herbicdes in forest resource management. J. For. 73(6):329-336.

Newton, M. 1977. Vegetation management alternatives in the practice of silviculture. Soc. Amer. Foresters Nat. Conv., Proc. p. 270-277.

Newton, M. and L. A. Norris. 1976. Evaluating short- and long-term effects of herbicdes on forest and range biota. Down to Earth 32(3):18-26.

Smith, D. M. 1962. The practice of silviculture. 7th edition, John Wiley.

Walker, C. M. et al. 1973. Rehabilitation of forest land. Journal of Forestry 71(3):136-162.

Walstad, J. D. 1976. Weed control for better southern pine management. Weyerhaeuser Forestry Paper No. 15. 44 p.

Weed Science Society of America. 1979. Herbicide handbook. 4th edition, Univ. of Illinois, Dept. Agron., Champaign.

CHAPTER 3

FOREST INSECT MANAGEMENT

A discussion of insect control methodology should start with a clear statement of what is really meant by *pest management.* The terminology has been accepted by both professionals and practitioners yet there are major differences in interpretation of the term. Some advocates propose that pest management would tend to eliminate chemical control. This is unlikely in managing most species even though the benefit would be welcomed. We define pest management as the total system of manipulating both hosts and insects that provides minimum losses with a minimum of total input.

When we boil down the various definitions of pest management to bare essentials, we discover that the major thought left is that the persons applying practices to reduce pest damage must *know what they are doing.* The pest manager must know the requirements of the pest and its total relationship to its environment, and must know both the ecological and economic consequences of actions taken. There is no room in pest management for simply killing the "bugs" whenever they become numerous or of automatically treating as near as possible to a particular calendar date no matter how many insects may be present. Pest management calls for melding various control tactics based on a general knowledge of the forest ecosystem, and a combination of strategies in a systems approach to management of the problem. Those who desire to study the concepts of pest management and the systems approach will find the book by Metcalf and Luckman to be a good review.

The Insect Problem

Insects are the most numerous living creatures on the earth. These animals have been very successful in evolution so that at present just over fifty percent of the species of living things are insects; about 72 percent of all animals are insects. The estimated number of over 704,000 species of insects in the world of which 290,000 are beetles *(Coleoptera)* is believed to be conservative.

Insects have been on the earth for a very long time; the first wingless forms in the fossil record are from the Devonian period and the first winged forms are from the Carboniferous. Most of the modern insect orders were developed by the Permian period. The insects were well established some 300 million years before man evolved.

The reasons for this discussion is to emphasize the great variations among insects and their adaptation to many habitats. The vast majority of them are terrestrial, though the numbers of aquatic forms (mainly fresh-water) are not small; the three percent figure for aquatics estimated by one author would mean over 21,000 species. The terrestrial forms are found in virtually every area of the world including the arid deserts. Numerous species are found in all forested areas of the world. The forest insects with which we are concerned here represent only a tiny minority of the numbers of species in the forest. The majority of forest insect species are highly beneficial to the total system.

The reader will find two publications very useful as manuals on the forest insects. These are *Western Forest Insects* and *Eastern Forest Insects.* Both manuals name and describe most of the significant insect problems of North American forests; references are given at the end of this chapter.

Forest Insects

We have stated that the "ultimate aim of forest entomology is to make possible the regulation, in the interest of man, of insect activities in forests and forest products." The concern in forest management is with the species which damage the forests we are attempting to grow and to the products produced. Apart from a desire to maintain them, we are less concerned with the thousands of beneficial species in the forest ecosystem which are essential but taken for granted. We will confine our comments to those beneficial species that directly affect the numbers of destructive pests. These are the insects commonly referred to as parasites, parasitoids and predators. Many of these are very important in the management of forest insect pest populations.

Some readers will have an interest in pest problems beyond those associated with the forest environment because these problems are not confined to the forested areas of our world. The same pests affect trees growing in non forest areas. Not only do insects in non-forest environments frequently meet with fewer enemies, but also we generally do not tolerate as much damage to high value shade trees as we do with the trees of the forest. Thus, the cost-benefit analysis changes and we may do things to control tree damaging pests in urban areas that we could not afford in the forest.

Determination of the amount of damage that may be tolerated in the forest is always difficult. The impact of forest insects is continually being studied in order to develop better thresholds of allowable damage. Mortality might be considered a basic threshold but even then we cannot always be sure. A certain amount of mortality in a forest stand might well serve as an appropriate thinning practice leading to greater growth of the remaining trees. Thus the knowledge that some trees are likely to die is not a suitable criterion to initiate control measures. We also need to know how many may die and how those deaths will be distributed. In contrast, the likely death of one high value shade tree is usually reason for decisive action. The forest ecosystem has a tendency to rebound in a recovery process. The urban shade tree is an individual entity with non-transferrable values. This ability to rebound from damage is an important factor in estimating thresholds.

Economic Injury Levels

The concepts of economic injury levels and economic thresholds have been presented by several scientists. These terms may be used most effectively in relation to direct (chemical) control, as preventive methods must be applied continuously and long before the economic threshold is reached.

The first of these terms (economic injury level) is the minimum amount of pest damage that will cause serious economic losses. It might be considered as the level of damage to the trees that can no longer be accepted. Here again the ultimate use of the trees is critical. The economic injury level in a plantation of Christmas trees is much lower than in a plantation managed for timber production.

The economic threshold of a pest population is always lower than the economic injury level. This is the pest density at which control should be applied to prevent the increasing pest population from reaching the economic injury level.

Figure I is a diagrammatic representation of these ideas. The general equilibrium position is the average population density of the pest over an extended time period. The location of this theoretical line in relationship to the economic threshold will vary according to species. Forest insects may be fitted to at least four categories in relationship to the economic threshold.

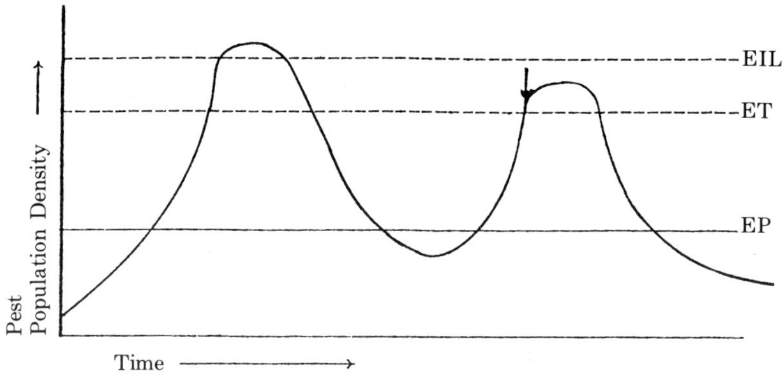

Fig. 3-1. Curves of population densities showing both natural and man-induced fluctuations. EP, equilibrium position; ET, economic threshold; EIL, economic injury level. Arrowhead indicates when control was applied.

1. Some very common forest insects may never occur in sufficient numbers to warrant control measures. The equilibrium position is far below the economic threshold and the variation from year to year is minimal. These species are so numerous in the forest that mention of specific examples is not meaningful.

2. A second group might be classified as those which occassionally cause losses. The economic injury level is exceeded on rare occasions and at higher levels intervention may be needed. Forest insects such as the pine butterfly, *Neophasia menapia;* the pandora moth, *Coloradia pandora* and the organ striped oakworm, *Anisota senatoria,* can be listed as among the many species in this category.

3. A third group is similar to #2 but may remain at or near the economic injury level over an extended period of years. This situation seems to result from a shift in the equilibrium position brought on by a change in the general environment such as the development of susceptible forest conditions. Some very

serious pests are included in this category: the spruce budworm, *Choristoneura fumiferana;* gypsy moth, *Lymantria dispar;* forest tent caterpillar, *Malacosoma disstria,* and others. Many of the more aggressive bark beetles such as mountain pine beetle, *Dendroctonus ponderosae;* and southern pine beetle, *D. frontalis,* are usually included in this group.

4. There are a few insect species whose impacts are always severe. The equilibrium position for them is above the economic injury level. An example is the white pine weevil, *Pissodes strobi.*

The reader can immediately see many loopholes in a classification such as that presented above. The categories overlap and parameters cannot be specifically defined. Nevertheless, this classification is a good basis for discussion because it clearly points out the wide range of evaluations which must be made. The economic injury level decreases as the value of the crop increases; for example, damage by a cone moth such as the southern pine coneworm, *Dioryctria amatella,* might be in category one in the forest but in a seed orchard could be close to category four because of the high values. Thus, economic injury level is a flexible concept which under the best of circumstances is difficult to measure accurately.

Use of Cultural Methods

Much has been said through the years about the use of silviculture in the control of forest insects and the use of other preventive procedures to protect forest products. Silviculture is the prime method for the prevention of losses. Silvicultural practices should be the leading component of any pest management scheme. Such methods are highly developed in Europe where preventive techniques are a part of most management prescriptions. A person on the ground hardly knows that insect prevention is one of the objectives of the practices applied. A similar approach is also common in the silvicultural systems used in the Western Pine Region of the United States.

In North America it is difficult to convince managers that the use of preventive methods is worth the investment. We have still not moved far enough away from extensive forestry to have developed the planning and investment required for full scale intensive culture of forests. There is still a tendency to avoid investment and to take the risks of greater losses as a part of forestry practice. We still prefer to risk an outbreak with the chance of high losses or a large scale and expensive control project in preference to a known lower cost for preventive work. Fortunately this attitude is gradually changing.

The report of the forest study team of the National Academy of Sciences states that all manipulations of stands designed to *maintain the overall health and vigor of the trees* may be a part of "cultural" control. Included in their definitions are such strategies as "avoidance," identification of "hazard zones" and the use of "sanitation" procedures.

Furniss and Carolin in *Western Forest Insects* point out that silvicultural control offers almost unlimited possibilities. Many of their suggestions relate to care in management and to the advantages of an additional investment in long range planning. We quote:

"In older forests, much can be done to lessen insect damage by avoiding injury to the trees from forest fires and other weakening influences; by keeping forests in a healthful condition through disposal of windfalls, slash, and other insect-breeding material; and by selective cutting operations to regulate composition and density and to remove trees most susceptible to insect attack. In plantations, consideration should be given to the selection of the site, the planting of species and varieties of trees best adapted to it, and proper spacing."

If these practices and others were adopted, nothing spectacular would happen but when the next rotation had been concluded losses would be substantially lower than if they were not applied.

General agreement by various authors on silvicultural methods lead to several generalizations. These have been expressed variously as follows:

1. Assure maintenance of stand vigor.
2. Maintain abundance of tree species and mixture of age classes where such mixtures normally occur.
3. Do not plant species not suited to soil or climate.
4. Do not permit trees to become overmature or decadent.
5. Develop the species mixtures found in the climax or subclimax forests of the region and site.

These principles summarize the major thrusts of a silvicultural control program. They have been known, essentially as stated, for more than twenty-five years, and we expect that they will be little changed in the future. The reason, of course, is that each of them is based on commonly observed forest behavior. However, the techniques used in the application of these principles has changed greatly in the same period and will continue to change in the future. The application of the principles is a complex and difficult process and advances in techniques will be welcomed. We must be careful about over-simplifying an analysis of a problem or its causes, and we can expect varying opinions while finding solutions. We will find no best method in silvicultural control but will continually adapt to differing pests, sites, environmental conditions, and management objectives.

Application

The general principles of silvicultural methods for control of forest insects will minimize damage if applied in a long-term program. We might question why such methods have not been used more widely if they work. The answer lies mainly in the intensity of management in our forests. Most of the recommendations require a high intensity of management which has not commonly been utilized over an extended period in North America. We already know that they are methods useful not only in intensive management but also as prevention for insect damage. The increasing value of managed forests promises their widespread application in the future.

Bark Beetles

Bark beetles are some of the most destructive tree-killing forest insects. With a few notable exceptions, they are generally secondary in that they attack only trees that are dying or are of low vigor or dying. Many species infest wind felled trees, in which large populations may develop, other species seem to be attracted first to overmature trees in poor health.

One of the earliest silvicultural control efforts was against the western pine beetle, *Dendroctonus brevicomis.* Frederick Keen developed tree classifications for ponderosa pine, *Pinus ponderosa,* for prevention of losses. His classification was based upon tree vigor and age. Generally speaking, the oldest trees with the poorest crowns were the most likely to be attacked. Keen's classification led directly to the risk-rating system that is currently applied in the field. The field methods became known as the sanitation salvage method which is implemented by removing trees of high risk before they are attacked. The method has been extremely successful with this insect species. Unfortunately attempts to apply the method to other species of bark beetles have not been as successful.

The mountain pine beetle, *Dendroctonus ponderosae,* also infests ponderosa pine and attacks several other species of pines as well. It is well documented that this insect also attacks trees under stress though the relationship to overmature trees is not like that of the western pine beetle. Sartwell and Stevens presented evidence that silviculture may be effective in preventing losses of ponderosa pine. They recommend early thinning to reduce damage because the severity of damage is greatest in dense young stands whose vigor has been reduced by severe overcrowding or stagnation. This method will work but is not applied on a significant scale because of the low intensity of management actually practiced in the region. It is likely that the very destructive outbreaks of this insect may become historic items in commercial forests once preventive methods are applied on a wide scale.

Lodgepole pine, *Pinus contorta,* is also infested by the mountain pine beetle. In this case a different management process is needed. High beetle hazard is age related and management is designed to harvest the stands before the trees reach the susceptible age.

These examples suggest that bark beetles could be kept in check indefinitely by cultural methods. Such may be the case when our management methods are more fully developed. However, stress on trees and low vigor may develop from a variety of sources. Either extreme drought or excessive moisture may create problems in some areas. Severe storms may cause excessive wind fall and breakage with resultant increases in populations of Douglas-fir beetle, *Dendroctonus pseudotsugae,* and spruce beetle, *Dendroctonus rufipennis.* Even the engraver beetles, *Ips* spp., may become destructive at times following drought and natural disturbances. We recommend that managers maintain stands in as vigorous condition as possible in order to prevent bark beetle problems as well as increase productivity. Any natural disasters should be followed by immediate salvage of damaged live trees in which beetles might breed.

Defoliators

There are many insect defoliators for which there are no feasible cultural control methods other than elimination of the host species. However, in all cases there is the possibility of isolation of hosts, e.g. by selective cutting or thinning, so that fewer may be available. This approach is effective if other tree species of equal value can be grown. In some regions such changes in stand composition is not practical.

It is unlikely that a silvicultural method could be developed for an insect like the pine butterfly or the elm spanworm, *Ennomos subsignarius.* These pests are rare so populations only occasionally develop into outbreaks. Other methods are more promising in such cases.

The many species of defoliating insects exist in differing ecosystems and are present because of special adaptations. There are species which may be affected by silvicultural operations. The spruce budworm, *Choristoneura fumiferana,* is an example that reaches outbreak population densities at intervals. Often an outbreak causes high mortality of balsam fir, *Abies balsamea,* and a considerable additional loss among associated spruce, *Picea* spp., trees. The development of an outbreak is closely timed to the ecological development of balsam fir. Outbreaks develop when two basic requirements are met. First, there is a need for an ample acreage of forested stands containing a high proportion of mature balsam fir. Second, a specific weather pattern must develop consisting of 3 or 4 years of favorable conditions for budworm development.

Recommendations for control of the spruce budworm have been proposed since the early twenties. A major requirement is the removal of vulnerable stands of balsam fir and managing such stands on short cutting cycles. The industry until very recently has been handicapped by an over supply of mature stands and insufficient markets for the trees. Under those circumstances it is impossible to develop a preventive control system throughout a region. Management of individual stands may be ineffective because in a serious outbreak the budworm may spread through all stands whether managed or untended. Removal of balsam fir early in the life of a mixed forest offers one option for managing the hosts; the consequences of this approach are unknown.

Past management practices in hardwood forests may make stands more susceptible to several defoliators or other insects. Selective cutting may favor the development of single species forests, especially when cutting is fairly light. The resultant single species stands may be more subject to damage by defoliators than the mixture of species that originally occupied the site. The silvicultural recommendation that seems most practical is to increase the cutting intensity to maintain the mixed species composition. Maintenance of a valuable mixture may reduce the economic impact of a specific pest, and reduces the likelihood of such a buildup.

Plantation Management

One other situation should be discussed in this brief review of the use of silviculture to prevent losses from forest insect populations. Our management activities involve a considerable attention to the planting of trees, generally to reforest areas but also to assure a plentiful supply of certain tree species. We have often failed to properly plan or allocate resources to this activity and sometimes we have carried out planting operations where we know failure would be likely. Frequently our zeal to reach goals gets in the way of common sense and the result is disastrous. Thus, the first rule in preventing losses to forest insects in plantations is to avoid planting trees on sites where there is little likelihood of success or where severe stress is likely. Forest insects and diseases are often blamed for losses under such circumstances when in actuality the planting should never have been done, or if done, it should have used different stock.

Planting trees on sites unsuited to them is not the only insect-related problem in plantation management. Before citing other examples, we need to mention a widespread failure in plantation management. We are all aware of the high cost of producing trees in a forest nursery, and yet that cost is small relative to the price of site preparation and planting. The land manager pays very heavily to prepare the site and to get the trees in the ground. It is a mystery of human behavior to observe this very large investment preceded by poor planning and followed by apparent abandonment. The common assumption seems to be that once planted, we only need to watch it grow. Even that may be giving too much credit, as evidenced when one tries to find some of the plantations established twenty to thirty years ago.

Current practices are not as bad as they were thirty years ago but there is still an attitude of "trust to luck" and avoid further costs. Plantations are especially susceptible to forest pest problems and the costs of preventing losses and to control insects or disease should be planned before the plantation is established. It might not be unreasonable to suggest that at least 10 percent of the cost of planting stock and establishment be reserved for such pest control activities. The reserve should probably be even higher for plantations of high cost stock that are expected to furnish especially valuable products.

The white pine weevil, *Pissodes strobi,* is a pest of conifers (pines and spruces) from coast to coast. This pest is especially significant in plantations, though the damage may be as severe in open grown natural regeneration. Damage to the "pasture-pines" in the east is clear evidence of what may happen.

This pest attacks the very best of the plantation trees. Those that are growing fastest and appear to be the best choices for crop trees are the first to be selected. The weevil problem has been a difficult one for managers from the time of earliest white pine plantings and there is still no clear or fully accepted way to prevent damage or control the problem. However, there are some preventive steps such as mixed plantings that have some effect, and direct control is possible. The 10 percent reserve mentioned earlier might be sufficient to control the pest, especially on better sites. As in so many other situations, this problem can be addressed if it is planned for and something done when needed.

The Saratoga splittlebug, *Aphrophora saratogensis,* is an example of another type of problem that may be prevented or even controlled by silvicultural operations. This sapsucking insect infests pine plantations most severely in locations where there is an abundance of sweet fern. The adults of this species feed on pines and other conifers, but the nymphal stages are found on plants in the understory. Such a pest may be eliminated by removal of the alternative host or by not planting in locations where the alternative host is abundant. We reiterate, that if a plantation is to be established where an alternative host is abundant the manager should include a reserve for controlling the alternative host or plan to live with the pest for the life of the stand.

Problem Areas

The discussion to this point has been generally related to the problem of insects in forests managed for timber production with some attention to problems in plantations established for future production. There are many other special cases where attention to problems caused by insects may mean the difference between success or failure in a business or a necessary land use plan.

Products Insects

The insects of concern to the manufacturer and consumers

of forest products are of concern from the time the trees are cut to long after the products have been put in service. Some of the products may be affected by insects which were present in the trees at the time of felling but we will not be concerned with those here.

Trees felled in the spring and summer are very quickly infested by insects which feed within the cambium/phloem area and in the wood itself. Bark beetles such as those of the genus *Ips* do not significantly alter the strength of products because they confine their activity to the phloem area and do not penetrate into the wood. However, bark beetles may be the reason behind another problem because they do bring into the tree the spores of bluestain fungi which cause discoloration of the wood. Treatment to kill the trees with the herbicide MSMA before felling will partially eliminate this problem in some northern or western pines, western hemlock and Douglas-fir.

Also invading the phloem area are insects that feed both in that area and, usually later in the cycle, in the xylem tissues. Many of these insects are beetles from the families Cerambycidae and Buprestidae; adults are commonly called long-horned beetles and metallic beetles while the larvae are known as round-headed and flat-headed borers, respectively. These insects may cause fairly serious losses in lumber grade if logs lie in the forest unattended for several months. Peeling or chemical debarking eliminates these losses.

A third group of insects attacks the felled trees and immediately invade the wood. Many of these are in the group called ambrosia beetles. These beetles commonly feed upon fungi that they raise within the extensive galleries they excavate in the wood.

Any of these problems may become serious but usually losses can be prevented or greatly reduced by careful planning. Logs should not be left in the forest any longer than absolutely necessary and when brought to the mill yard should be protected from further infestation. This is often done with a water treatment but may involve the use of chemicals. In Europe, logs are peeled when felled to prevent infestation. All of these insects attack through the bark; any treatment that destroys the bark as an attractive entry or resting point will reduce or eliminate damage.

After the lumber is manufactured, the problems with insects are not over – much lumber may be destroyed by powder post beetles. *Lyctus* beetles lay eggs in the pores of hardwoods and, in time, larvae may reduce the interior of a board to powder. Kiln drying will destroy such pests and the use of protective paints and varnishes will prevent further infestation. Termites may also attack wood in use and will feed upon most wood species. Special precautions are needed to prevent entry of drywood termites into wood structures; metal or chemical barriers to termite movement should be utilized. Dryness is effective prevention for wet-wood termites. Wood preservatives may be used to prevent infestations in wood in direct contact with soil.

Preservatives are also used to protect wood used in the ocean where infestation by marine borers may be extremely destructive. These pests are mollusks rather than insects but are mentioned here because protection of wood from their effects is done in a similar manner to that for insects.

Shade Tree Insects

Many of the pests of ornamental trees are the same species as those in the forest but there is a significant difference. First, the trees are of much greater value than a forest tree of the same species and, second, the individual tree along a street or in a yard is somewhat out of place and is more visible than when growing in the forest. Thus, problems which might go unattended in the forest may be serious on a high value tree in an urban location.

The manager or pest control operator will have many additional insect pests to deal with in the urban situation. These include not only those which go unnoticed in the forest but also those which attack other urban plants not found in a native forest environment.

In the forest, the high costs involved and the low economic margin means that the use of direct control is avoided as much as possible. In contrast, the high value of ornamental trees may result in intensive use of chemicals to "protect" the trees. Intensive use, however, should not be confused with excessive use. There are many excesses in urban pest control. In view of opportunities for human exposure, it is imperative that these be kept to a minimum.

Seed Orchards

Seed orchards are becoming more common as tree plantations expand. The long range objective is to produce seed from genetically selected stock to improve the productivity of our forests through higher growth rates and improved wood quality. Each tree in a seed orchard is of very high value and the orchard manager must be concerned with the protection of each individual and its crops. A discussion of these problems as related to seed orchards in the South has been presented by southern scientists.

The problem with pests in seed orchards is not confined to the health of the orchard tree but also to the quantity and quality of seeds produced. Each year a large portion of the seed from forest trees is destroyed by insects. During most good seed years there is an excessive amount of seed produced in a forest and ample quantities are present to regenerate the stands. But, even in the forest, there are times when seed is scarce.

No seed orchard produces a surplus of premium seed. Thus, every seed produced is of high value and losses must be prevented. Coneworms, *Dioryctria* spp.; seedworms, *Laspeyresia* spp.; seed chalcids, *Megastigmus* spp.; and acorn weevils, *Curculio* spp., are among the many species of significance. Few of these are amenable simply to cultural treatment.

Sugar Bush Management

This special use of the forest is of limited significance nationally but of great concern to those directly involved. Insects are not usually serious problems in well managed sugar bush stands because in such stands the trees are properly spaced to assure vigor and rapid growth. However, managers occasionally have problems with outbreaks of defoliators such as the saddled prominent, *Hemerocampa guttivitta,* or the forest tent caterpillar, *Malacosoma disstria,* and wood borers, especially the sugar maple borer, *Glycobius speciosus.* Here again the high value of the individual trees may require that control be undertaken. The problem may be in management; if such is the case thinning is advisable to improve the overall vigor of the sugar bush as well as to improve the productivity of the stand. Many years are required to develop a productive sugar bush, thus good planning, prompt recognition of problems, and careful management are vital.

Christmas Trees

The production of Christmas trees requires specialized management activities and a considerable investment during the 5 to 10 years before returns are realized. A wide variety of tree species are grown, although the numbers of any given species may be relatively small within specific regions. Chemicals are used throughout the rotation for a variety of reasons, herbicides are utilized effectively in managing the planted areas and fertilizers are often required to correct nutritional deficiencies.

Management of Christmas trees is typically very intensive. Managers are continually on the watch for any increase in the activities of insects and diseases. Any insect problem which destroys the foliage or causes mortality of branches may make a tree useless for the market. Growers tend to use more insecticides and other chemicals because of these problems. A grower producing trees on a fairly large scale should have a private applicators license and should own equipment suitable for treating the trees.

Insect problems are varied. A manager in the Lake States could have serious losses caused by feeding of European pine sawfly, *Neodiprion sertifer,* on scotch pine. In another area and on a different tree species the problem might be the yellow headed spruce sawfly, *Pikonema alaskensis.* Other growers might have problems with the white pine weevil on pines and spruce or the Nantucket pine tip moth, *Rhyacionia frustrana,* on pines. Balsam fir may be severely injured by the spruce budworm, *Choristoneura fumiferana,* or the balsam gall midge, *Paradiplosis tumifex.* A grower might have problems with the latter two insects in the same year or with some other problem such as those caused by the balsam twig aphid, *Mindaris abietinus,* which causes curled and twisted needles on the new shoots.

We have illustrated this problem area by mentioning only a few of the many insects that may cause a grower difficulty. The owner who is serious about raising Christmas trees should talk to other growers about their problems and seek advice from the sources available to everyone. An illustration of the type of materials available to growers is in the Extension Bulletin titled "Christmas Tree Insect Management."

Wind Break Management

A windbreak is defined as a planting of trees and shrubs designed to reduce or eliminate undesirable effects of strong winds. Trees are selected carefully for these purposes but they are often planted in treeless areas far from their natural habitat. These windbreaks are very effective and one wonders why more are not planted. Many windbreaks are established every year but unfortunately nearly half of these fail to survive in a satisfactory manner. These specialized stands of trees and shrubs must be protected from livestock, weeds, rodents, insects, disease and fire; often they are ignored.

In a study in North Dakota there were six kinds of insects injurious to the most common species of trees and shrubs. High populations of these insects were associated with deficient rainfall, temperature extremes, desiccating winds, and unfavorable periods.

The six insect problems in North Dakota were:

1. Caragana blister beetle, *Epicauta subglabra* (Fall)
2. Woolly elm aphid, *Eriosoma americanum* (Riley)
3. Boxelder twig borer, *Proteoteras willingana* (Kearfott)
4. Poplar petiole gall aphid, *Pemphigus populitransversus* (Riley)
5. Fall cankerworm, *Alsophila pometaria* (Harris)
6. Spur-throated grasshoppers (several species)

The problem insects in another region could be entirely different. A prudent manager will be alert for damage, however, and will prescribe in each case according to need and specific characteristics of the pest and its life cycle.

Recreation Areas

The problems associated with the management of forested recreation sites are similar to those of the person managing timber producing areas with a few notable exceptions. Generally the same insects are problems but the trees in a recreation site may be far more valuable for aesthetic and comfort reasons. The timber manager might accept losses of timber in isolated areas but proportional losses of stands around recreation areas would be unacceptable. Recreationists may also be bothered by biting and sucking flies and so require special attention by the manager.

It is interesting that sometimes a pest that is really of little significance to the forest may have a very great impact on recreational use of the area. People sometimes just do not like to see large numbers of insect larvae. Many people left the area infested by the red humped oakworm, *Symerista canicosta,* because they could not stand seeing thousands of these caterpillars crawling over picnic tables and in about every location imaginable (Figure 3-2).

Land Reclamation

Finally, we should comment briefly on the problem of land

*Figure 3-2. Larvae of the red-humped oakworm *Symerista canicosta* on the stem of an oak in Michigan (photo by T. C. Eiber).

reclamation associated with strip mining or erosion problems. Site conditions are often very severe and the plants chosen may be under great stress. Insect problems under such circumstances may be more frequent than usual. Some of the best species for reclamation such as black locust may be severely damaged by insects. The locust borer, *Megacyllene robiniae,* has been a very serious pest of planted black locust trees.

Types of Damage

We have discussed some of the problems with insects in forests, how these problems may be reduced through attention to management practices and some of the specialized situations where damage by insects may be of significance to the land manager. We need now to summarize some of the insect problems in relationship to the type of damage caused. A broad classification of insect pests will facilitate discussion of the

Figure 3-3. Severe defoliation of Douglas-fir caused by the Douglas-fir tussock moth, *Orgyia pseudotsugata* (U.S. Forest Service photo).

various problems a land manager may need to solve.

1. Leaf-eating insects
2. Sucking insects
3. Insects of shoots and roots
4. Cone and seed insects
5. Bark beetles
6. Other phloem insects
7. Wood destroyers and wood products insects

Leaf Eating Insects

The foliage of trees furnishes food for many thousands of insect species. Among them are some of the more serious pests of our forests, such as the spruce budworm, *C. fumiferana,* which may destroy trees over vast acreages. Outbreaks seem to last for years and eventual death of the host trees may be the only way in which the populations finally crash. There are also insects like the Douglas-fir tussock moth, *Orgyia pseudotsugata,* which also causes much mortality after severe defoliation (Fig. 3-3). Outbreaks of the latter seem to be short in duration, and factors other than death of the host apparently influence the collapse of populations.

The general effect of defoliation varies greatly according to the species of host tree, the health of the host, the habits of the feeding insect and the time of year in which the defoliation

Figure 3-4. Larvae of the Virginia pine sawfly, *Neodiprion pratti pratti,* feeding upon pine needles (U.S. Forest Service photo).

takes place. Conifers, if fully defoliated, are often more seriously injured than deciduous trees because of their inability to refoliate. Hardwoods may survive three or more successive seasons of severe defoliation while conifers may die following one. However, there is no hard and fast rule and the manager should anticipate losses of growth and some mortality after any severe defoliation. There may also be a lag effect with damage and losses occurring many years after the insect outbreak has passed.

There are several types of defoliators; (1) *leaf miners* feed between the surface layers of leaves, (2) *skeletonizers* remove the soft tissues leaving the veins and one surface intact, (3) *shot hole feeders* chew holes in the leaves and (4) *wholeleaf feeders* consume all leaf tissues as they feed. Some of these groups are rather difficult to control whether with chemicals or by other means. Leaf miners, for example, are protected from chemicals due to their location within the leaf tissues. Some species may be skeletonizers in their early larval stages but when larger they consume entire leaves of the trees.

Most of the serious pests that consume leaves are in two orders of insects, the Lepidoptera and Hymenoptera. The first are the moths and butterflies while the second includes the defoliating insects known as sawflies, Fig. 3-4. Some of the more serious leaf eating insects are included in Table 3-1.

Table 3-1. The leaf eating insect pests of North America – some important species, their hosts and significance in forestry.

SPECIES	HOSTS	LOCALITY	SIGNIFICANCE
Western black-headed budworm, *Acleris gloverana*	Western hemlock, sitka spruce, true firs	Northern Calif. to the Yukon and Alaska	Low
Fall cankerworm *Alsophila pometaria*	Elm, apple, oak, other hardwoods	Maritimes to Alberta and south in U.S.	Low to moderate
Orange-striped oakworm, *Anisota senatoria*	Oaks	Canada to Georgia west to Texas	Low
Birch skeletonizer, *Buculatrix canadensisella*	Paper birch, other birches	Southern Canada South to North Carolina	Low
Large aspen tortrix, *Choristoneura conflictana*	Aspens, balsam, poplar, willow	Canada and U.S. coast to coast	Low
Spruce budworm, *Choristoneura fumiferana*	Balsam fir, spruce	Canada and eastern USA	High
Western spruce budworm, *Choristoneura occidentalis*	Douglas-fir and true firs	Canada and western USA	High
Jack pine budworm, *Choristoneura pinus*	Jack pine	Lake States and Canada	Moderate
Larch casebearer, *Coleophora laricella*	Tamarack, western larch	Southern Canada northern USA	Moderate to high
Pandora moth, *Coloradia pandora*	Ponderosa, coulter and lodgepole pines	Rocky Mountains West	Moderate
Walnut caterpillar, *Datana integerrima*	Many deciduous esp. walnut, butternut, pecan and hickory	Southern Ontario eastern USA	Moderate
Saddled prominent, *Hemerocampa guttivitta*	Beech, paper birch, sugar maple and	Southeast Canada and Eastern USA	Moderate to high
Fall webworm, *Hyphantria cunea*	More than 100 species	United States and southern Canada	Moderate

Table 3-1 (Continued)

SPECIES	HOSTS	LOCALITY	SIGNIFICANCE
Hemlock looper, *Lambdina fiscellaria fiscellaria*	Balsam fir, white spruce and hemlock	Newfoundland south to Georgia	High
Western hemlock looper, *Lambdina fiscellaria lugubrosa*	Western hemlock and coniferous associates	Oregon to Alaska	High
Satin moth, *Leucoma salicis*	Poplars and willow	New England and eastern Canada west to Washington and Oregon	Low
Aspen blotchminer, *Lithocolletis tremuloidiella*	Aspens	California to Alberta	Low
Gypsy moth, *Lymantria dispar*	Oak, birches, other deciduous trees	Northeastern USA west to Michigan	High
Western tent caterpillar, *Malacosoma californicum*	Many species of deciduous woody plants	Western Canada and United States	Moderate to low
Forest tent caterpillar, *Malacosoma disstria*	Hardwoods - aspens maples, gums	United States and Canada widespread	Moderate to high
Lodgepole sawfly, *Neodiprion burkei*	Lodgepole pine	Montana and Wyoming	Low
Red-headed pine sawfly, *Neodiprion lecontei*	Jack, red, short-leaf, loblolly	Southeastern Canada and eastern USA	High
Jack pine sawfly *Neodiprion pratti banksianae*	Jack pine, also red and scotch	Southeastern Canada to Lake States	Moderate
European pine sawfly, *Neodiprion sertifer*	Scotch, red, jack, Mugho pines	Much of eastern U.S. southeastern Canada	Low to moderate
Pine butterly, *Neophasia menapia*	Ponderosa pine	Northwestern USA and Canada	High

Table 3-1. Continued

SPECIES	HOSTS	LOCALITY	SIGNIFICANCE
Mourning cloak butterly or spiny-elm caterpillar, *Nymphalis antiopa*	Elm, willow, poplar, hackberry others	Subarctic, North America, widespread	Low
Douglas-fir tussock moth, *Orgyia pseudotsugata*	Douglas-fir, true firs	British Columbia, Rocky Mountains to coast	High
Yellow-headed spruce sawfly, *Pikonema alaskensis*	White, black, red, blue, Norway, Engelmann spruce	Widespread where spruce grows	Low
Larch sawfly, *Pristiphora erichsonii*	Tamarack, western larch, other larches	Canada, Alaska all of northern USA	Moderate to high

The defoliators described in Table 1 are only a partial listing of species the manager may find. No defoliators were included from other orders though there are notable examples such as the Walkingsticks (Orthoptera) and the leaf beetles (Coleoptera). The manager should be always alert to these leaf eating insects because, as with many of the serious pests, losses may occur quickly giving very little time to take action. The early activities of the insects in a developing outbreak may go unnoticed. Some of the new trapping methods and attractants (pheromones) provide improved early detection of defoliator outbreaks such as the Douglas-fir tussock moth, *O. pseudotsugata*.

Sucking Insects

The insects discussed in the previous section have chewing mouthparts with which they eat the leaf tissues. The sucking insects have mouthparts developed for piercing the tissues to reach the plant fluids within the cells. Most of these insects are in the two insect orders Hemiptera and Homoptera. The effect on trees may be inconspicuous; only a few species are tree killers but others may cause significant losses in growth or appearance. Sucking insects may also vector disease-causing organisms and generally reduce tree health. These insects are seldom of great significance in forests but may be very damaging to shade trees and in plantations.

Table 3-2. The sucking insects of North America – some species commonly found on forest and shade trees.

SPECIES	HOSTS	LOCALITY	SIGNIFICANCE
Cooley spruce gall adelgid, *Adelges cooleyi*	White, blue, sitka, Engelmann spruce, Douglas-fir	Coast to coast in northern USA, southern Canada	Low
Balsam woolly adelgid, *Adelges piceae*	Balsam fir, fraser fir, other true firs	Southeastern Canada northeastern USA Pacific Northwest	High
Pine spittlebug, *Aphrophora parallela*	Scotch pine preferred; red, jack, slash, loblolly pines	Southern Canada eastern U.S.	Low
Western pine spittlebug, *Aphrophora permutata*	Pines, Douglas-fir, hemlock, spruce, fir	Western states southern Canada	Low
Saratoga spittlebug, *Aphrophora saratogensis*	Red, jack, pitch, loblolly pines	Southeastern Canada and Eastern U.S.	Moderate
Pine needle scale, *Chionaspis pinifoliae*	All species of pine	Widespread in North America	Moderate
White pine aphid, *Cinara strobi*	White pine	NE, Lake States to Carolinas	Low
Sycamore lace bug, *Corythucha ciliata*	Sycamore; also ash, hickory	Eastern U.S. and southern Canada to Colorado and New Mexico	Low
Beech scale, *Cryptococcus fagi*	Beech	Southeastern Canada south in eastern U.S.	High
Oystershell scale, *Lepidosaphes ulmi*	Over 100 species of hardwood	Southern Canada and all of U.S.	Moderate
Pinyon needlescale, *Matsucoccus acalyptus*	Pinyon and foxtail pine	Southwestern USA	Moderate
Red pine scale, *Matsucoccus resinosae*	Red pine	Northeast	High
Balsam twig aphid, *Mindarus abietinus*	True firs	Transcontinental	Low to moderate

Some of the commonly encountered species on trees are noted in Table 3-2.

Insects of Shoots and Roots

Many insects feed upon the meristematic tissues of the tree. These are the areas of the plant where growth is taking place and includes the growing tips of the shoots and roots as well as the cambium area beneath the bark. The insects that feed upon the shoots and roots of trees include some well known and difficult to manage problems. The majority of the insects in this group are either the tip and shoot moths (Lepidoptera) or beetles (Coleoptera).

Many of these insects cause deformities rather than mortality of trees, though infestations in small trees may cause death. These insects include both primary and secondary problems; a primary insect is capable of living and growing in a healthy tree. Secondary insects are not able to complete development in healthy hosts. The terms sound simple enough but in practical situations there are so many factors influencing the hosts and the insects that the differences are sometimes difficult to evaluate.

Possibly the best known of these insects affecting terminals is the white pine weevil or sitka spruce weevil, *Pissodes strobi.* This insect can cause very severe deformities (Fig. 3-5) in plantations and in open grown natural stands. It is the major pest of eastern white pine and the reason that white pine plantations are not more common. This insect has prevented intensive, large-scale management of this valuable, fast-growing eastern pine. Some examples of insects of the terminals of trees (shoots and roots) are presented in Table 3-3.

Table 3-3. The shoot and root insects of North America – some examples and the localities in which they are common.

SPECIES	HOSTS	LOCALITY	SIGNIFICANCE
Twig pruner, *Elaphidionoides villosus*	Oaks, hickory, maple, locust, walnut, elm and other hardwoods	All of eastern states, southern Canada	Low to moderate
Western pineshoot borer, *Eucosma sonomana*	Ponderosa pine, jack pine	British Columbia western states east to Lake States	Low
Cottonwood twig borer, *Gypsonoma haimbachiana*	Poplars	Ontario, northern tier of states to Oklahoma and Texas	Low

Table 3-3. Continued

SPECIES	HOSTS	LOCALITY	SIGNIFICANCE
Pales weevil, *Hylobius pales*	Young pines of most species	Ontario, eastern and southern U.S., west to Great Plains	Moderate
Pine root collar weevil, *Hylobius radicis*	Scots, jack, red, pitch and white pines	Southern Canada, Northeast and North Central	Moderate locally
Northern pitch twig moth, *Petrova albicapitana*	Jack, scotch and lodgepole pines	East to west in southern Canada, south to Colorado and to Carolinas	Low
White grubs, *Phyllophaga* spp.	Roots of most trees	Throughout North America	Low to moderate
White pine weevil, Sitka spruce weevil, *Pissodes strobi*	White pine, jack pine, Norway and sitka spruce other pines and spruces	Coast to Coast	High
European pine shoot moth, *Rhyacionia buoliana*	Scotch, red, ponderosa, white, jack, pitch and other pines	Southern Canada, northeastern states and Pacific Northwest	Moderate to high
Western pine tip moth, *Rhyacionia bushnelli*	Ponderosa and monterey pine	Montana east to Nebraska and southwest	Low to moderate
Nantucket pine tip moth, *Rhyacionia frustrana*	Nearly all species of pine in range	Eastern, central and southern states	Moderate

Cone and Seed Insects

In managed forests reproduced without planting, cone and seed insects generally are unnoticed and economically unimportant. Attention has been focused more closely on these insects during recent years because of the development of seed orchards. These seed producing areas represent a high investment in selected growing stock, and loss of seed is not acceptable on a large scale.

Seed production areas have also been developed in some sections of the country where stands of exceptional quality are present. These stands may require special treatment to control pests. Cone losses as high as 60 to 70 percent between flowering and maturing are not acceptable. Some examples of the variety of insects involved are presented in Table 3-4.

Table 3-4. Seed and Cone insects of North America – examples of the various genera and species attacking forests.

SPECIES	HOSTS	LOCALITY	SIGNIFICANCE
Douglas-fir cone moths, *Barbara colfaxiana*	Douglas-fir and true firs	California to British Columbia to Colorado	Low
White pine cone beetle, *Conophthorus coniperda*	White pine	Eastern North America range of white pine	Low
Ponderosa pine cone beetle, *Conophthorus ponderosae*	Ponderosa and jeffrey pine	Washington to California and southwest	Low
An oak curculio, *Conotrachelus naso*	16 species of oaks	Maine to Florida southern Canada west to Rockies	Low
Douglas-fir cone midge, *Contarinia oregonensis*	Douglas-fir	California to British Columbia	Low
Pecan weevil, *Curculio caryae*	Pecan and hickory	New York to Iowa, south to Texas	Low to moderate
Filbert weevil, *Curculio occidentis*	Oaks and filberts	British Columbia to New Mexico	Low to moderate
Fir coneworm, *Dioryctria abietivorella*	Many conifers, esp. firs and Douglas-fir	Transcontinental	Low
Southern coneworm, *Dioryctria amatella*	Longleaf, lob-lolly, slash, and short-leaf pines	Virginia to Florida west to Texas	Low
Shortleaf coneborer, *Eucosma cocana*	Shortleaf and other pines	Massachusetts to Florida	Low
Slash pine flower thrips, *Gnophothrips fuscus*	Slash pine, jack pine, other pines	Southern and eastern states	Low
Hickory shuckworm, *Laspeyresia (Cydia) caryana*	Hickory and pecan	Southern Canada to Texas	Low

Table 3-4. Continued

SPECIES	HOSTS	LOCALITY	SIGNIFICANCE
Spruce seed moth, *Laspeyresia (Cydia) youngana*	Cones of spruces	Western states, Canada to Alaska	Low
Balsam fir seed chalcid, *Megastigmus specularis*	Balsam fir	Canada and eastern USA	Low
Douglas-fir seed chalcid, *Megastigmus spermotrophus*	Douglas-fir	Western U.S. and Canada-range of Douglas-fir	Low

Figure 3-5. Typical damage resulting from attack by white pine weevil, *Pissodes strobi* (photo by F. B. Knight).

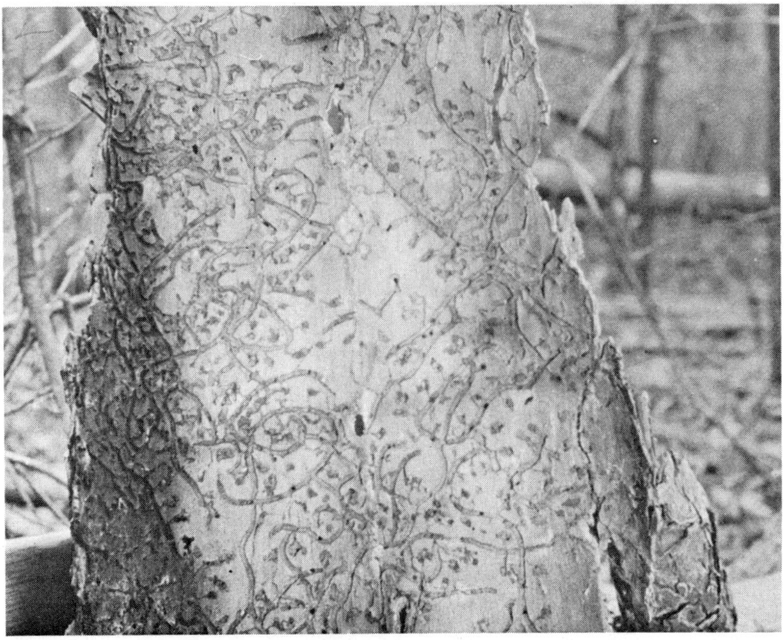

Figure 3-6. Characteristic gallery patterns of the Southern pine beetle, *Dendroctonus frontalis* (U.S. Forest Service photo).

Bark Beetles

Bark beetles are included in the family Scolytidae. Among them are some of the most destructive insects in North America. Direct mortality of mature trees is greater from bark beetles than from any other natural cause when comparisons are made on a year to year basis. Each year bark beetles kill trees in outbreaks somewhere in the country. The genus *Dendroctonus* contains the majority of the more destructive species, though there are other serious pests in the genera *Ips* and *Scolytus*. Some of these such as the smaller European elm beetle, *Scolytus multistriatus,* may carry the causal agents for very serious tree diseases (Dutch elm disease).

Each genus in the family has a characteristic manner of working (Fig. 3-6) and there is also much variation among individual species. All species excavate egg galleries in fresh phloem and the larvae feed through the succulent tissues of the inner bark until fully grown. Most bark beetles may be classified as secondary insects because they typically attack trees in poor health; some species are capable of developing to maturity in trees that

Figure 3-7. Characteristic pattern of attack by the mountain pine beetle, *Dendroctonus ponderosae*, on ponderosa pine (photo by F. B. Knight).

were apparently healthy when first attacked. Thus, some species may become primary insects during epidemics. These insects apparently succeed in killing trees through the process of a massive attack on individual trees.

Control of bark beetles is especially difficult when outbreaks are in progress. Stands may be managed so that such outbreaks may be avoided and it is possible through use of salvage, pheromones and some chemicals, to reduce the intensity of infestations. The forest manager should act quickly to salvage losses early in an outbreak to prevent a build up to severe levels (Fig. 3-7) in managed stands and to maintain the quality of the product from the dead trees. Bark beetles are presented in Table 3-5, all *Dendroctonus* are included, and examples of other genera are listed.

Table 3-5. The bark beetles of North America – some important species, their hosts and their significance to land managers.

SPECIES	HOSTS	LOCALITY	SIGNIFICANCE
Round headed pine beetle, *Dendroctonus adjunctus*	Ponderosa, limber, chihuahua pines	Colorado and Utah south to Mexico	Moderate
Larger Mexican pine beetle, *Dendroctonus approximatus*	Ponderosa and Mexican pines	Colorado and Utah south to Mexico	Low
Western pine beetle, *Dendroctonus brevicomis*	Ponderosa and coulter pine	British Columbia to New Mexico	High
Southern pine beetle, *Dendroctonus frontalis*	Southern pines Mexican pines	Southeast, southwest, Mexico	High
Jeffrey pine beetle, *Dendroctonus jeffreyi*	Jeffrey pine	California	Moderate
Lodgepole pine beetle, *Dendroctonus murrayanae*	Lodgepole and jack pine	British Columbia to Utah and Colorado	Low
Mountain pine beetle, *Dendroctonus ponderosae*	Ponderosa, lodgepole, western white, limber, coulter and other pines	British Columbia to northern Mexico	High
Douglas-fir beetle, *Dendroctonus pseudotsugae*	Douglas-fir and western larch	British Columbia to California	Moderate to high
Allegheny spruce beetle, *Dendroctonus punctatus*	White spruce and sitka spruce	Transcontinental far North	Low
Spruce beetle, *Dendroctonus rufipennis*	Spruces	Transcontinental, south to Rocky Mtns. and New Mexico	Moderate to high
Eastern larch beetle, *Dendroctonus simplex*	Tamarack	Eastern U.S. and Canada west to Alaska	Low
Red turpentine beetle, *Dendroctonus valens*	All pine species plus spruce and larch	Northeast U.S. across Canada into Mexico	Low to moderate
Black turpentine beetle, *Dendroctonus terebrans*	Southern pines and red pine	Southern New England to Texas	Moderate

Table 3-5. Continued

SPECIES	HOSTS	LOCALITY	SIGNIFICANCE
Birch bark beetle, *Dryocoetes betulae*	Birch, beech, sweetgum, wild cherry	Transcontinental in Canada, south to Florida and Mississippi	Low
Native elm bark beetle, *Hylurgopinus rufipes*	Elms	Transcontinental in Canada range of elms	Moderate
Sixspined ips, *Ips calligraphus*	Many pine species	Transcontinental in Canada and south to Texas	Moderate
Arizona fivespined ips, *Ips lecontei*	Ponderosa pine	Southwest into Mexico	Moderate
California five-spined ips, *Ips paraconfusus*	Many pine species	Oregon to southern California	Moderate
Pine engraver, *Ips pini*	Spruces and pines	Boreal forests of North America south to Tennessee	Low to moderate
Eastern ash bark beetle, *Leperisinus aculeatus*	Several ash species	Southeastern Canada and eastern U.S.	Low
Western cedar bark beetle, *Phloeosinus punctatus*	Western red-cedar, juniper	California to British Columbia, Utah	Low
Balsam fir bark beetle, *Pityokteines sparsus*	Balsam fir	Eastern Canada and eastern USA, balsam fir range	Low
Foureyed spruce beetle, *Polygraphus rufipennis*	Spruces	Transcontinental Canada and northern USA	Low
Smaller European elm bark beetle, *Scolytus multistriatus*	Elms	United States and southern Canada	High
Hickory bark beetle, *Scolytus quadrispinosus*	Hickories, pecan and butternut	Quebec to Georgia west to Oklahoma and Texas	Low
Fir engraver, *Scolytus ventralis*	True firs	Western U.S. and southern Canada	Low

Other Phloem Insects

There are numerous other insects which feed in the phloem of trees. Some of these may spend all of their larval life in these tissues while others develop for a time in the phloem then later bore into the xylem where they pupate. Very few of these bore through the bark in their attack, instead they commonly lay eggs in bark crevices and the larvae bore into the phloem from that location. Thus, the adults of these species do not construct egg galleries. Many of these insects are beetles and commonly of two families, Cerambycidae (Fig. 3-8) and Buprestidae. The larvae of these two families are easily distinguished as the former are called roundheaded borers while the latter have a distinctly flattened prothorax and are called flatheaded borers.

Another group of phloem borers are moths, the majority of which belong to two families – the Cossidae and the Sesiidae. The carpenterworm, *Prionoxystus robiniae,* is especially destructive (Fig. 3-9). This pest invades scars on healthy hardwoods and may be the cause of very large losses in value of hardwood lumber. Many species of both the *Lepidoptera* and Coleoptera are secondary in their habits and under certain circumstances would be considered beneficial rather than destructive. Some examples of the species in the group are presented in Table 3-6.

Table 3-6. Phloem insects common in the forests of North America – specific examples from some of the genera.

SPECIES	HOSTS	LOCALITY	SIGNIFICANCE
Ponderosa pine bark borer, *Acanthocinus princeps*	Ponderosa and other pine species	British Columbia to California	Low
Bronze birch borer, *Agrilus anxius*	Paper, yellow and other birches	Range of birch in Canada and U.S.	Moderate
Twolined chestnut borer, *Agrilus bilineatus*	Chestnut, oaks, other hardwoods	Southern Canada eastern U.S.	Moderate
Pacific flatheaded borer, *Chrysobothris mali*	Many hardwood species	Western states and Canada	Low
Pecan carpenterworm, *Cossula magnifica*	Pecan, oaks, and hickories	Southern United States	Moderate
Red oak borer, *Enaphalodes rufulus*	Oaks	Southern Canada to southeast and west to Texas	Moderate

Table 3-6. Continued

SPECIES	HOSTS	LOCALITY	SIGNIFICANCE
Sugar maple borer, *Glycobius speciosus*	Sugar maple	Southern Canada to southeast and west to Texas	Moderate
Locust borer, *Megacyllene robiniae*	Black locust	Canada and United States where black locust grows	Moderate to high
California flatheaded borer, *Melanophila californica*	Ponderosa and jeffrey pines plus other pines to a lesser extent	Pacific north, west to California and	Low to moderate
Hemlock borer, *Melanophila fulvoguttata*	Hemlock and other conifers	Eastern U.S. and Canada	Low
Whitespotted sawyer, *Monochamus scutellatus*	Spruce, pine, Douglas-fir, true firs	Transcontinental U.S. to Alaska	Low
Ash borer, *Podosesia syringiae fraxini*	Various ash species	Texas to midwest southern Canada	Low
Carpenterworm, *Prionoxystus robiniae*	Oaks, elms, poplars, other hardwoods	United States and eastern Canada	High
Ribbed pine borer, *Rhagium inquisitor*	Pines and other conifers	Transcontinental U.S. and Canada	Low
Poplar borer, *Saperda calcarata*	Poplars and willows	United States and Canada	Low
Maple callus borer, *Sylvora acerni*	Sugar and red maple	Southern Canada eastern U.S. south to Florida	Low
Western cedar borer, *Trachykele blondeli*	Western red cedar, Cupressus and others	British Columbia to California to New Mexico	Low
Sequoia pitch moth, *Vespamima sequoiae*	Pines, and Douglas-fir	California to British Columbia, Idaho, Montana	Low
Rustic borer, *Xylotrechus colonus*	Most hardwood species	Eastern United States	Low

Figure 3-8. The poplar borer, *Saperda calcarata,* is one of the long horned beetles. The larvae are round headed borers (U.S. Forest Service photo).

Wood Destroyers and Wood Products Insects

The wood destroyers include a wide variety of insects, Buprestids and Cerambycids similar to the phloem borers, horntails, ambrosia beetles, powderpost beetles, carpenter ants, and termites. Some of these attack recently dead trees or logs while others attack wood that has been in use for years. Control of these insects may be difficult to accomplish, thus, the best recommendation is proper attention to prevention of damage. Most of the damage to wood in use can be prevented through use of paint and preservatives.

The most destructive of the group are the termites which can be controlled using chemicals. Damage may be prevented by using chemical barriers or by following specified codes for building construction in localities where termites are a problem. Ambrosia beetles are serious problems in recently logged areas where logs cannot be removed immediately (Fig. 3-10). Damage may be reduced by careful planning of operations and protecting or debarking logs after removal.

Some of the species of wood destroying insects are included in Table 3-7.

Table 3-7. Insects that destroy wood – examples from various groups.

SPECIES	HOSTS	LOCALITY	SIGNIFICANCE
Furniture beetle, *Anobium punctatum*	Mostly old hardwood materials	Canada and northeast U.S.	Low
Golden buprestid, *Buprestis aurulenta*	Ponderosa pine and Douglas-fir	Pacific Northwest to British Columbia	Moderate
Black carpenter ant, *Camponotus pennsylvanicus*	Most tree species	Eastern U.S. and Canada	Low to moderate
Columbian timber beetle, *Corthylus columbianus*	Maples, sycamore, yellow poplar, boxelder, birches, and others	Northern tier of states Georgia and Arkansas	Moderate
Old-house borer, *Hylotrupes bajulus*	Dry coniferous wood	Massachusetts to Florida, west to Texas	Moderate
Western drywood termite, *Incisitermes minor*	Dry wood of any species	California, southwest to Mexico	Moderate
Southern lyctus beetle, *Lyctus planicollis*	Oak, ash, hickories	Widely distributed most serious in southern states	High
Ambrosia beetle, *Platypus flavicornis*	Pines and other species	Canada to southern states	Low
Cottonwood borer, *Plectrodera scalator*	Cottonwood and willows	Southern states	Low
Eastern subterranean termite, *Reticulitermes flavipes*	Wood in contact with soil	Southern states north to Massachusetts and Ontario	High
Leadcable borer, *Scobicia declivis*	Oak, maples and other seasoned hardwoods	California and Oregon	Low
Blue horntail, *Sirex cyaneus*	Spruces and pines	Transcontinental	Low
Pigeon tremex, *Tremex columba*	Many hardwood species	Transcontinental	Low
Striped ambrosia beetles, *Typodendron lineatum*	Firs, spruces, pines, Douglas-fir	Canada and northern states	Low to moderate

Figure 3-9. This gallery of the carpenterworm, *Pronxystus robiniae,* was formed in the living tree thirty years before the photo was taken (U.S. Forest Service photo).

Figure 3-10. Ambrosia beetles attack felled trees. Note the piles of white powder which is an indication of heavy infestation in this log (U.S. Forest Service photo).

Insecticides and Their Application

We have discussed various aspects of the insect problems that forest managers may need to consider, and in this treatment we have emphasized host and pest management. The manager must attempt to solve problems without use of direct control techniques when possible. However, there will be occasions when the insect populations become so destructive that some direct method will be needed. The number of insect species that might be involved are numerous as evidenced by the tables of insect examples in the previous section of this chapter. Those with high significance have caused either severe problems over large areas or they have been a significant deterrent to management or utilization of forests and timber products. The ones listed as low significance to forest management or wood utilization may be highly significant to someone managing for a special product, or by someone interested in maintaining ornamental plantings. Any of the insects named could require direct action. The major weapons available for direct control of insects are insecticides though other methods may be used on a smaller scale.

Insecticidal Materials

Chemicals have been used for insect control for centuries. Homer referred to use of sulphur at or about 1000 B.C. and the Chinese were using arsenic in control of garden insects before 900 A.D. Comments on such uses are common in the literature and the constant struggle against pests continues. Humans have done very poorly through the centuries in the battle against insects; in ancient times the number of weapons was pitifully few. As the numbers of people increased, the need for more efficient control also grew and by the middle 1800's a number of botanical materials and inorganic poisons were in common use. The first of the synthetic organic materials were marketed in Germany in about 1892 but it was the Second World War that provided the greatest stimulus for laboratory manufactured organic insecticides.

Chemicals for insect control must be properly formulated and diluted before they can be effectively and efficiently used against insects. Undiluted materials are expensive and often highly toxic to plants and animals. However, when properly diluted the risk to other organisms is greatly reduced and the insect pest can be controlled at the least cost. Formulations of both insecticides and herbicides are discussed in detail in Chapter 4.

Inorganic insecticides were at one time important in control of forest insects but were never used on a wide scale. Some of the common materials were calcium arsenate, lead arsenate, cryolite, sulfur compounds, and mercury compounds. Lime sulfur is still used for the control of certain scale insects and mites. Even though these inorganic compounds are not currently used in forest insect control we should be aware that at some future date they might be found useful.

As we continue this discussion it will be evident that there are many types of insecticides useful for different purposes. Some are stomach poisons, meaning that the insects must eat the poison, while others are contact poisons which will penetrate the surface of the insect. Fumigants volatize readily and reach the insect through its breathing process. Such materials are effective in places where the gases cannot escape. Still other specialized types of chemicals are classified as systemic insecticides. These are chemicals which are absorbed by the plant and translocated to the structure on which the insect may be feeding. Systemics have definite advantages though they are not presently applicable to many forest insect problems.

Organic Insecticides

Most of the presently registered insecticides are synthetic organic materials; a few of these compounds are plant derivatives (botanicals) such as nicotine (not used in forestry), pyrethrum, and rotenone. Most of the heavily used materials are of the former group.

Pyrethrum is obtained principally from the plant *Chrysanthemum cinerariaefolium* and is made up of four toxic esters as follows:

Pyrethrin I,	$C_{21}H_{38}O_3$
Pyrethrin II,	$C_{22}H_{28}O_5$
Cinerin I,	$C_{20}H_{28}O_3$
Cinerin II,	$C_{21}H_{28}O_5$

The cinerins are more stable than the pyrethrins and cinerin I and pyrethrin I seem to be more toxic than the other two components. The materials are expensive and are very susceptible to photodecomposition. However, the pyrethrins are valuable insecticides because of their low mammalian toxicity and rapid insecticidal action. They are also effective on a broad range of insects and degrade rapidly into innocuous compounds.

Research has been concentrated on lowering costs and solving the problem of rapid breakdown in sunlight. This has resulted in the development of synthetic pyrethroids. The first of these with the characteristic high activity against insects and low mammalian toxicity but with greatly increased stability was described in 1973. Experiments since that time have resulted in additional compounds and field tests have been promising. Photostable pyrethroids will evidently be more expensive on a weight basis compared to other chemical insecticides because synthesis is more complex. However, the rates of application may be considerably lower because of the higher toxicity to insects. These materials may provide some of the future weapons against forest insects; they are being tested currently.

Chlorinated hydrocarbons (organochlorine compounds) were the first large group of synthetic organic insecticides. The most famous of these was synthesized in 1874 by a German chemist but it was not recognized as an insecticide until World War II. DDT is now rarely used in the United States because of adverse side effects in the environment caused by its fat solubility and slow degradation rate. However, the material did serve well in the early years following World War II as an effective control for most leaf eating insects.

There are many chlorinated hydrocarbons still in use. Two examples of significance in forestry are methoxychlor and lindane. Their chemical names and chemical structures will serve as examples for the group.

Common Name: Lindane

Chemical Name: γ-1,2,3,4,5,6-Hexachlorocyclohexane

Structure:

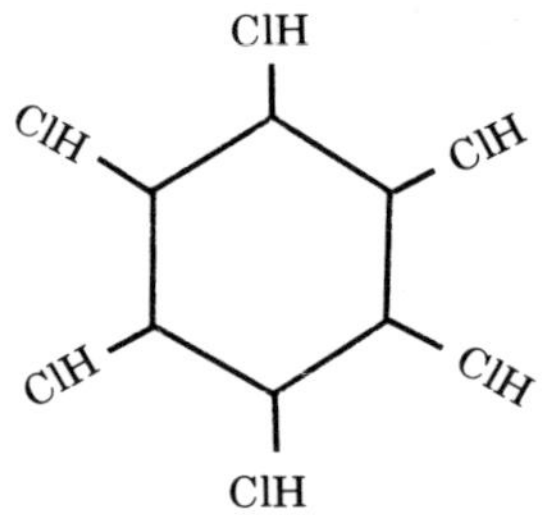

Common Name: Methoxychlor

Chemical Name: 2,2-Bis(p-methoxyphenyl)-1,1,1-trichloroethane (88%) and related compounds

Structure:

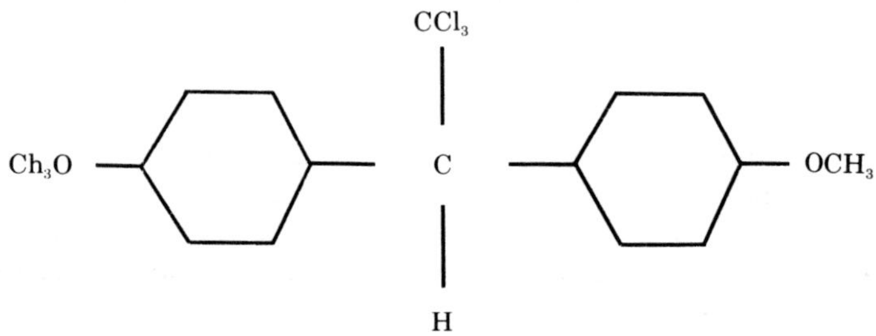

Lindane is more volatile than DDT and has been used effectively in control of bark beetles. Residues only last a few days at high temperatures and exposure to sunlight. Methoxychlor is less toxic than DDT to warm-blooded animals, does not accumulate as rapidly in animal fat and thus is much more acceptable.

Organophosphorus compounds are a second group of materials including several of significance to foresters. Parathion, the first organophosphorus compound synthesized for use as an insecticide was developed in Germany in 1943. Many of these compounds, including parathion, are very toxic to mammals, including people. Early research then was directed toward finding compounds that are highly toxic to insects but of low mammalian toxicity. Malathion was the first of these and is very effective against many forest insects. Some members of this group are contact poisons while others are systemics.

The two examples we have selected are malathion and phosphamidon; both are contact materials and are safe to handle. Malathion has very low mammlain toxicity and is considered one of the safest organophosphorus insecticides. Phosphamidon is one of the class of systemic insecticides that has very low toxicity to fish, thus is particularly suitable for wooded regions close to lakes or rivers.

Common Name: Malathion

Chemical Name: 0,0-dimethyl phosphorodithioate of diethylmercapto-succinate

Structure:

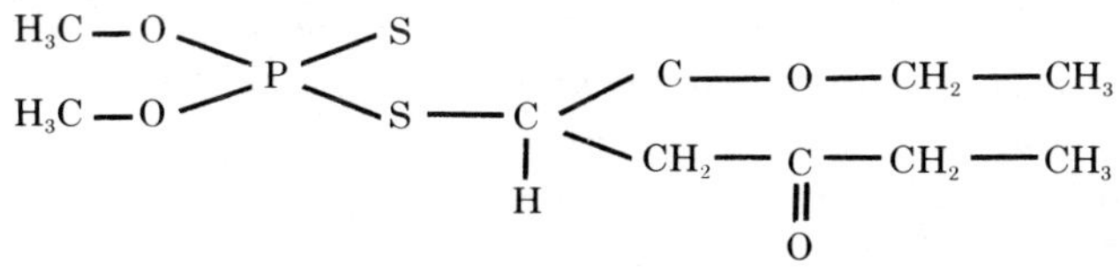

Common Name: Phosphamidon

Chemical Name: 0-[2-chloro-2-(diethyl-carbamoyl)-1-methylvinyl]-0-0-dimethyl phosphate

Structure:

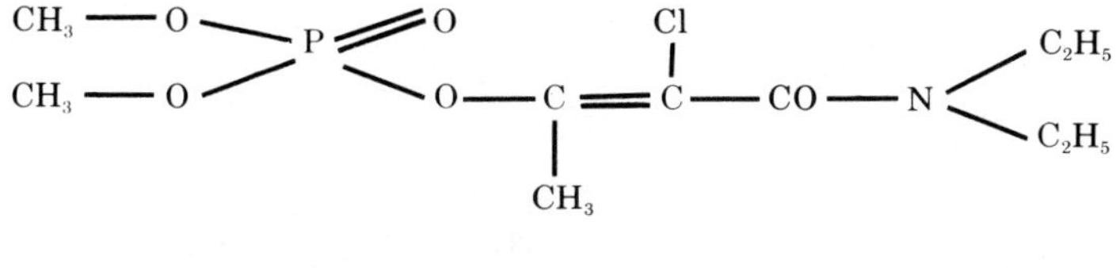

Carbamates were introduced later than the other groups but, because of their insecticidal properties, several compounds have been accepted. Carbaryl is the most widely used carbamate and is registered for several forest insects including the spruce budworms. It has low mammalian toxicity and is effective against a broad spectrum of insects.

Common Name: Carbaryl
Chemical Name: l-Naphthyl-N-methylcarbamate
Structure:

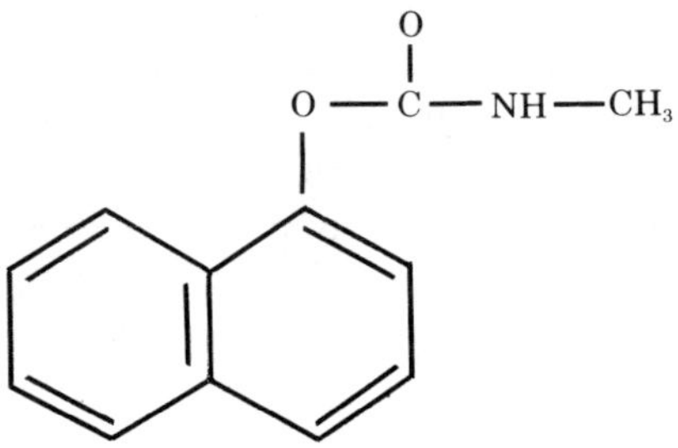

Fumigants are considered here as a separate group though some of them are similar chemically to compounds already discussed. These are usually used in enclosed spaces, and in forestry are used mainly in nurseries or on forest products. An exception would be the treatment of colonies of the Texas leafcutting ant, *Atta texana,* in the forest environment.

Common Name: Methyl bromide
Chemical Name: Monobromomethane
Structure: CH_3Br

Common Name: Para-dichlorobenzene
Chemical Name: 1,4-Dichlorobenzene
Structure:

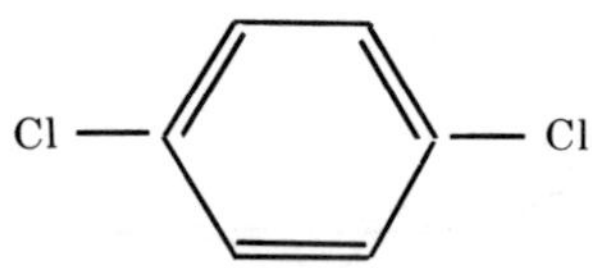

We have not attempted to identify all of the organic insecticides in this statement. The major requirement for the manager is to use procedures described elsewhere in this book in selecting the appropriate registered material and applying the insecticides according to label instructions. Some of the insecticides used on forest insect problems are presented in Table 3-8. Toxicological data are presented in Table 5-1.

Table 3-8. Insecticides utilized in control of insects affecting forests, nurseries, ornamental trees and forest products.

OP = Organophosphate, CH = Chlorinated Hydrocarbon, C = Carbamate, F = Fumigant, G = Growth Regulator

Common Name	Other Designation	Type
acephate	ORTHENE[R]	OP
aminocarb	MATACIL[R]	C
azinphosmethyl	GUTHION[R]	OP
carbaryl	SEVIN[R]	C
chloropicrin	PICFUME[R]	F
chlorpyrifos	DURSBAN[R]	OP
demeton	SYSTOX[R]	OP
diazinon	SPECTRACIDE[R]	OP
difluorobenzuron	DIMILIN[R]	G
o-dichlorobenzene	ORTHODICHLOROBENZENE	F
p-dichlorobenzene	PARADICHLOROBENZENE	F
diflubenzuron	DIMILIN[R]	G
dimethoate	CYGON[R]	OP
disulfoton	DI-SYSTON[R]	OP
ethylene dibromide	EDB	F
endosulfan	THIODAN[R]	CH
fenitrothion	SUMITHION[R] ACCOTHION[R]	OP
gardona		OP
imidan		OP
lime sulfur		Inorganic
lindane	γBHC	CH
malathion	CYTHION[R]	OP
methomyl	LANNATE[R]	C
methoxychlor	MARLATE[R]	CH
methyl bromide		F
mirex		CH
naled	DIBROM[R]	OP
oxydemeton-methyl	METASYSTOX-R	OP
pentac		CH
phorate	THIMET[R]	OP
phosphamidon	DIMECRON[R]	OP
proposur	BAYGON	C
pyrethrins		Botanicals
rotenone		Botanicals
thimet		OP
toxaphene		CH
trichlorfon	DYLOX[R]	OP

Other insect controls

Most of the materials already described act as poisons; however, one organic chemical in Table 3-8 is a growth regulator.

Common Name: Difluorobenzuron
Chemical Name: N-[4-(Chlorophenzylaminocarbonyl)]-2,6-difluorobenzamide
Structure:

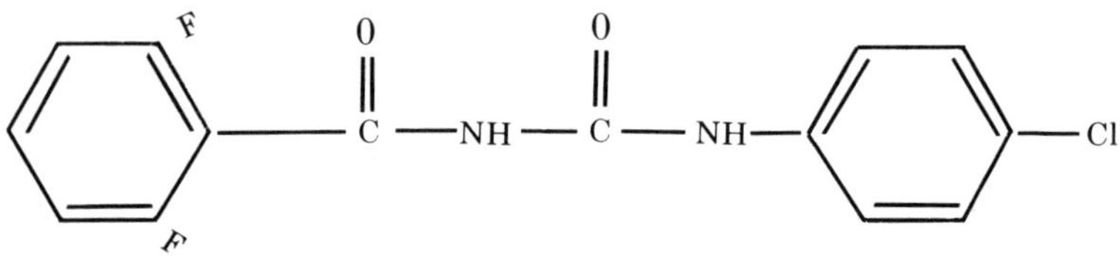

Insect Growth Regulators (IGR's are currently being tested as insecticides and as a group are commonly referred to as third generation materials. These materials (IGR's) have greater selectivity than the chemicals previously discussed and evidently have fewer undesirable side effects.

Abnormal morphology is the first and most readily observed aspect of IGR action on insects, and it is caused by an interference with normal biochemical processes. Two main types of growth hormones have been tested; the juvenile hormone analogues which mimic the insect's juvenile hormone and the molting disruptants that prevent the insect from molting. Currently the latter are prime candidates for forest insect control.

Molting disruptants are effective at amazingly low concentrations. They are stable in sunlight and have a residual life on foliage of over three weeks. Since they do upset the molting process they kill target insects in the early instars before much damage occurs. Difluorobenzuron was the first of these molt inhibitors to be developed but others are now being tested. It has given good results in tests against Douglas-fir tussock moth, gypsy moth, and spruce budworm. The material and/or some of the others will eventually be registered for use on forest insects.

A variety of pathogenic organisms attack insects and some have a very high degree of species-specificity. There are several examples of very effective insecticides that have been developed from bacterial and viral agents.

Bacillus thuringiensis (B.t.) is a commercially available microbial pesticide that has been registered for use against numerous insects. It was registered for the Douglas-fir tussock moth in early 1976. The nucleopolyhedrosis virus of the Douglas-fir tussock moth has also been tested intensively and has been registered. These virus materials are much more specific than B.t. and are considered to be some of the most effective and safest insecticides. It is not expected that such materials will be discovered for most serious pest problems and it may not be economically practical to produce them when they are developed. There are many insects not affected by any serious microbial disease. On the other hand, a broad spectrum material such as B.t. could be useful against many species.

There are several potentially important materials but only one other has been registered for use on a forest insect; the virus of the gypsy both. Methods have been developed for producing quantities of both of the registered viruses. Studies on the tussock moth indicate that the dry powder has a storage life of at least fifteen years. This means that these highly species-specific materials may be held after production until needed.

Sex pheromones are chemical substances produced and released by one sex to attract the opposite sex. Many of the Lepidoptera and Coleoptera that are serious forest pests produce these attractant pheromones. Some of these materials have been synthesized and are being used for survey purposes. Pheromones also have a potential for use in control by mass trapping techniques and confusion of populations through disruption of normal communication between the sexes, or as supplements to other integrated pest management procedures. Promising tests have been done with bark beetles and gypsy moth. None of these materials are registered currently for control of insects, but eventually this may happen.

The discussion of insecticides must include a comment on host-parasite relationships which may be altered considerably by the use of chemicals. There are records of successful control projects against an insect that have resulted in an outbreak of another pest. These stimulated outbreaks have generally been short in duration and are generally the result of the destruction of the insect parasites or predators of the secondary pest. A common example is the build up of mite populations following spraying of defoliators. The mites are maintained at low populations and are generally not pest problems except on ornamentals, but destruction of their predators can have serious consequences.

Insecticides can also have beneficial effects on parasite effectiveness in that an abundant parasite population may be forced to work on a reduced host population after spraying of an insecticide. This is especially true of materials that must be ingested by the host. This effect is a temporary one as parasite populations are governed by the availability of hosts. It is well to do what we can to avoid destruction of parasites and predators as well as other insects in an area.

Perhaps a few fundamental rules might be repeated at this time as they apply to this situation:

1. Always follow label instructions and never use more insecticide than is needed.
2. Be careful with timing of applications so that a minimum of disturbance to populations of other organisms will occur.
3. Use materials that are as specific to the host as possible even though material costs may be higher; stomach poisons and disease materials (bacteria and viruses) are examples.

Application

Any discussion of the use of insecticides must include some comments on *application,* even though the types of equipment are discussed elsewhere in this text. There is always a tendency in insect control to use a system of overkill. A conservative but decisive approach is the best in forest insect management. Here again we find a failure in application when managers confuse conservative methods with avoidance of action. We must realize that, in application, a garden sprinkler (Fig. 3-11) may still be the ideal instrument for a specific use in control rather than an expensive mechanical monster costing thousands of dollars. However, when an elaborate system is needed, a watering can is of little use.

The effectiveness of any spray program depends on certain basic application principles: the selection of the proper insecticide, proper formulation, application at the correct time, use of the right equipment, and exact attention to directions.

The large spectacular spray projects attract attention but most insect control campaigns are small scale operations. It is interesting to note that over 90 percent of all the acreage sprayed to control forest insects from aircraft have been against 3 species: spruce budworm, *Choristoneura fumiferana;* western spruce budworm, *Choristoneura occidentalis;* and the gypsy moth, *Lymantria dispar.* There have been other sizeable aerial spray projects against a number of species but each has been small when compared to operations against the three named.

Most of these large aerial spray programs have been regional in extent covering large acreages. They have been cooperative ventures between Federal, State and private organizations. The

Figure 3-11. The ordinary sprinkling can may be an effective piece of equipment for applying insecticides (photo by F. B. Knight).

Table 3-9. Insecticides registered in the United States for use on Forest Insects with some Examples of Species for which registrations Exist[1]

Material	Pest Examples
Acephate (Orthene)	Gypsy moth, spruce budworm, aphids, tent caterpillar, fall webworm
Azinphos-Methyl	Coneworms, aphids, Nantucket pine tip moth, spider mites
Bacillus thuringiensis	Pine butterfly, tent caterpillar, gypsy moth, Douglas-fir tussock moth
Carbaryl	Spruce budworm, gypsy moth, birch leafminer, European pine sawfly
Carbophenothion	European pine shoot moth, pine tip moths
Cacodylic acid	Douglas fir beetle, southern pine beetle, spruce beetle, pine engraver beetles
Carbofuran	Cone beetles, coneborers, coneworms
Diazinon	Bagworms, aphids, tent caterpillars, leafhoppers, loblolly pine sawfly
Dimethoate	Hemlock scale, aphids, fir cone midge, Nantucket pine tip moth.
Ethylene dibromide	California flatheaded borer, mountain pine beetle, spruce beetle, fir engraver beetle
Fenitrothion	Spruce budworm
Lindane	Bark beetles, white pine weevil, ash borer, pine root collar weevil
Malathion	Saratoga spittlebug, aphids, larch casebearer, spider mites
Methoxychlor	Elm bark beetles, tent caterpillars, leafhoppers, fall cankerworm
Nucleopolyhedrosis virus	Douglas-fir tussock moth, gypsy moth
Oxydemetonmethyl	Aphids, leafminers, spider mites
Trichlorfon (Dylox)	Gypsy moth, spruce budworm, bagworm, forest tent caterpillar

[1]Source: Guidelines for the Control of Insect and Mite Pests of Foods, Fibers, Feeds, Ornamentals, Livestock, Forests, and Forest Products. USDA Agriculture Handbook #554, 1979.

mobility of forest insects is such that small outbreaks are not common in many species. There are exceptions, of course, such as insects on ornamentals in seed orchards and plantations. The large group of problems associated with forest products are other exceptions to the outbreak idea. These may be confined to very specific situations. The practibility of using insecticides at any scale should be related to the scope of the outbreak and the mobility of the pest.

We cannot provide a complete list of registered insecticides along with a list of insects for which registration has been approved because the list would be long and generally incomplete or inaccurate. Table 3-9 provides some information on the registrations available in 1979 but this table is only useful as an indicator of the variety of registrations. The manager should contact a county agent, state forester or a consultant to obtain information on registered materials for a specific problem.

Chapter 3 References

Anderson, J. F. and H. K. Kaya (Editors). 1977. Perspectives in forest entomology. Academic Press, NY, 428 pp.

Anonymous. 1969. Principles of plant and animal pest control: Vol. 3. Insect-Pest Management and Control. National Academy of Sciences, Wash., D.C., 508 pp.

Anonymous. 1975. Pest control: An assessment of present and alternative technologies: Vol. 1. Contemporary Pest Control Practices and Prospects and Vol. IV Forest Pest Control. National Academy of Sciences, Wash., D.C., Vol. 1 506 pp, Vol. II 170 pp.

Baker, W. L. 1972. Eastern forest insects. U.S. Dept. Agr., Forest Service Misc. Publ. 1175, 642 pp.

Borror, D. J., D. M. Delong and C. A. Triplehorn. 1976. An introduction to the study of insects. 5th edition, Holt, Rinehart & Winston Inc., NY, 852 pp.

Brown, A. W. A. 1978. Ecology of pesticides. John Wiley & Sons, NY, 525 pp.

Caswell, R. L. (Editor). 1977. Pesticide handbook – Entoma. Entomol. Soc. of Am., College Park, MD, 246 pp.

Daly, H. V., J. T. Doyen and P. R. Ehrlich. 1978. Introduction to insect biology and diversity. McGraw Hill, NY, 564 pp.

Furniss, R. L. and V. M. Carolin. 1977. Western forest insects. U.S. Dept. Agric. Forest Service Misc. Publ. 1138, 654 pp.

Johnson, W. T. and H. H. Lyon. 1976. Insects that feed on trees and shrubs. Cornell Univ. Press, NY, 463 pp.

Knight, F. B. and H. J. Heikkenen. 1980. Priniciples of forest entomology. 5th edition, McGraw Hill, NY.

Metcalf, R. L. and W. H. Luckmann (Editors). 1975. Introduction to insect pest management. John Wiley & Sons, NY, 587 pp.

Prebble, M. L. (Editor). 1975. Aerial control of forest insects in Canada. Dept. of the Environment, Ottawa, Canada, 330 pp.

CHAPTER 4

APPLICATION TECHNOLOGY

A pesticide as it arrives in the hands of the user is still a long way from controlling insects or weeds. The active ingredients are normally used in very small amounts, yet they must be distributed in such a way that most of the offending pests are exposed. For the safety of non-target species, it is important that excessive amounts of chemical are avoided, and that harmful amounts are kept out of places where they do damage.

Herbicides and insecticides used in silvicultural practice are usually distributed in the form of sprays. Granular products, pellets, drenches and other forms may be used in the nursery or seed orchard. Virtually all products are diluted for application as sprays because of the difficulty of distributing very small volumes of liquids or solids uniformly. The way chemicals are diluted and subsequently handled plays a large role in results obtained and damage done to non-target species or off-site areas.

Baits are used in certain dry insecticide products. These highly specific preparations are prepared by mixing an attractive food substance with the toxicant, usually for placement in sheltered locations. Total amounts of toxicant prepared in this way are small.

Properties of a Spray Mix

Sprays released for insect and weed control must be designed to provide adequate coverage of a large amount of vegetative surface. When applied by aircraft, they must cover potential targets with very small volumes of material, i.e. between one pint and ten gallons per acre. For example, a gallon per acre must be spread in a film 1/25,000 inch (.001 mm) thick in order to cover only one layer of foliage completely. The average forest canopy is several layers deep, and includes much extra surface in stems, bottoms of leaves, and understory. If all surfaces were covered, the film would have to be less than 1/100,000 inch (.00025 mm) thick. The deposition of drops in a pattern that will give such a film entails the breakup of a fluid into many small drops, which must then be well distributed. A major technical accomplishment in forestry is the careful matching of fluid and delivery system.

Spray liquids vary from light oil solutions to thickened emulsions (Fig. 4-1). They may have one or several physical phases, including different kinds of liquid states and suspended solid particles. Each of these mixtures has different properties when forming a spray pattern.

Light oils have low surface tension, and tend to break up into very small droplets when injected into air at high speed. They are used when low volumes of liquid must be dispersed for maximum density of droplets on the target. Typically, oils of this kind are used for insecticides applied in low-volume and ultra-low volume applications; herbicides applied to bark must also be applied in oil. Diesel fuel is the most common of the light oils used as a spray carrier, but other, non-phytotoxic, oils are used as well. Chemicals used in oil are nearly always oil soluble, and oil sprays are nearly always clear solutions of one phase. Oil sprays adhere well to foliage, and spread into a thin film for excellent coverage and penetration. The tendency for small drops to spread on contact makes oil a useful solvent for applying insecticides in very low volumes.

Water is the most common carrier for herbicides, and also is used widely for insecticides. Water is heavier than oil, is more viscous than the lightest oils and has much higher surface tension than oil. It also has lower affinity for plant surfaces (adhesion). When applied through the same spray system as oil, it will form larger drops, and leave fewer droplets per square inch on the target. Viscosity, surface tension and adhesion are involved in this phenomenon.

Viscosity and surface tension of water are affected by adjuvants designed to modify mixtures for spraying. Surfactants reduce surface tension, causing drops to adhere to foliage and to spread into a film on the target. Emulsifiers with oily additives create two-phase mixtures, in which drops of one phase

Figure 4-1. Diagrams of spray mixtures having different physical properties.

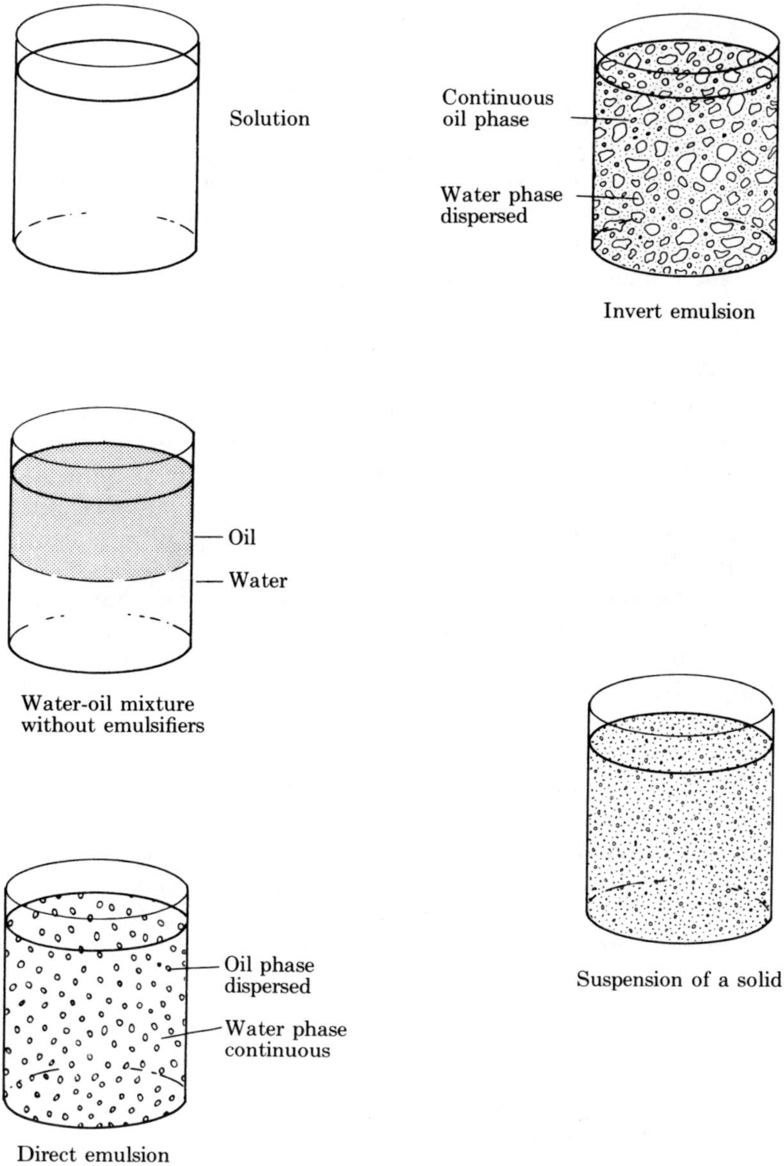

are dispersed in a vehicle of another phase. The oily components often provide better penetration into foliage and better spread of drops on leaves. When oil is added to water and dispersed as drops of oil suspended in water, the mixture is a direct emulsion, more commonly known simply as an *emulsion*. When the water is added to the oil phase in the presence of certain emulsifiers, drops of water are dispersed in oil to form an *invert emulsion*. A normal emulsion tends to be somewhat more viscous than water, but an invert emulsion may reach mayonnaise-like consistency needing special pumps, and nozzles with large orifices. These properties are useful in controlling droplet size, hence in managing drift potential and distribution on the target.

When insoluble solids must be applied in a spray, water is used as the carrier. Solids are prepared as *wettable powders* or as *flowable* liquid concentrates, which are finely ground solids dispersed in a thickened liquid matrix (see Chapter 2). These materials can be *suspended* in liquid with agitation, but settle out when agitation ceases. Most of the particles are of colloidal size, but some are large enough to clog very small nozzles, and must be used with appropriate combinations of fine filter screens and nozzles.

Solids do not normally affect the properties of the spray mixture greatly. Occasionally, the use of minimum volumes in aerial equipment requires very concentrated suspensions. When concentrated to more than one pound per U.S. gallon, the liquid may take on the properties of a slurry or light mud, in which case viscosity may be increased.

Most dry pesticide products are nearly insoluble in water, but there are exceptions. When a moderately soluble powder, such as hexazinone, goes into partial solution, there is some risk of crystallization when temperature drops, causing clogging of nozzles or screens. A few dry products, such as dalapon and certain amitrole preparations, are highly water soluble, and do not have this problem.

Formation of a Spray

On its way to becoming a spray, a mixture goes through several steps, each of which affects its usefulness as a pesticide deposit. The first is the mixing process. This is followed by storage in the spray tank, pumping to the nozzles, and finally, release into the air (by the nozzle) to form a spray.

The mixing process is important in determining uniformity

of concentration. Most pesticide products are heavier than water or oil. If not stirred adequately at time of mixing, they sink to the bottom of the tank, and may be drawn off in layers producing very strong mixtures at first, and ineffective sprays later. To avoid this problem, it is always recommended that mixing be done in a tank with plenty of agitation in it. Agitation is needed at the time of mixing and for a period before the mix is moved into the sprayer tank. Agitation is especially important for suspensions of wettable powders, which tend to form a "cake" at the bottom of the tank when allowed to stand. Clear solutions do not require continuous agitation, but should be thoroughly mixed while batching. A *true solution will not settle out after thorough mixing, but emulsions and suspensions will do so.* All need thorough mixing initially; emulsions and suspensions should be agitated continuously.

After placement in the spray tank, emulsions or suspensions should be agitated while spraying. Well dispersed emulsions, well agitated before placement in the tank and applied immediately after being put in the sprayer, may not require agitation in the sprayer or aircraft. Wettable powders always should be agitated while being applied. Hydraulic or mechanical agitation will maintain a good suspension if the material is well dispersed when placed in the tank. Mechanical agitation is usually needed for mixing suspensions. A solution mixed thoroughly before adding to the spray tank does not need further agitation. The distinction between solutions, emulsions and suspensions is extremely important in application technology.

Sprayers and Their Components

A sprayer is a device that physically delivers a volume of liquid to a target in dispersed droplets. Every sprayer contains a reservoir, a nozzle system, a pumping device and a more or less elaborate plumbing system.

The spray mix is moved from the reservoir tank to the nozzle by the pump which also supplies the pressure needed for atomization. Uniform pressure is essential to uniform nozzle performance. Pumps usually have more capacity than is needed. Pressure is kept from rising too high by a "pressure relief valve" in the line between pump and boom on-off control valve. The pressure relief valve allows the excess volume to flow back into the tank and bleeds off excess pressure. It also provides hydraulic or "by-pass" agitation through re-circulation.

Pumps vary in several important respects. *Piston* pumps are self-priming and are usually used for high pressure work, often with a hose-and-gun type sprayer. They are expensive and subject to excessive wear from abrasive effects of wettable powders or flowable chemicals. *Diaphragm* pumps are similar, but adapted to lower pressure. Abrasives have little effect on them. *Roller* or *vane* pumps are useful for medium pressures and volume deliveries. They are also self-priming, and subject to wear from wettable powders, but they are quite simple in construction and less costly to repair. *Centrifugal* pumps are not usually self-priming. They may be used for all kinds of spray materials without excess wear. They produce high volume delivery at low to medium pressure. They should not be used for invert emulsions because shearing action of the impeller affects the physical nature of the emulsion.

When the boom control valve is opened, the spray mix flows under pressure from the pump into the boom and then to a nozzle, of which there may be many. Enroute to the boom, the mixture goes through a filter or screen system in which particles larger than 1/100 inch or 1/50 inch, depending on screen, are removed from the liquid. These screens must be readily accessible for frequent cleaning, or they will clog and reduce boom pressure. They are an important part of the sprayer, however, because they prevent plugging of nozzles.

The nozzle is a highly engineered device. At the nozzle, the fluid moves through a channel to an orifice that projects a stream or spreading cone or sheet of liquid, which is broken into droplets as it spreads by the surface tension of the fluid. Rate of spread determines droplet size. As the cone or sheet spreads, it eventually becomes so thinly stretched that drops form along the edge furthest from the nozzle. The faster the film spreads, the smaller the drops. An increase in viscosity or decrease in pressure reduces the rate of expansion, and leads to the formation of larger drops. A whirlplate causes a cone-type nozzle to produce a wider cone angle, leading to rapid spread and, thus, fine breakup.

Nozzles may be designed to produce fine sprays, or to produce sprays with very few fine droplets. Injection of the spray mixture into *moving* air affects the formation of a spray. Orientation of the nozzle on an aircraft boom has an important added influence on droplet size through modification of wind action on the disintegrating liquid stream.

Different application jobs require different nozzles. Ground sprayers use the following kinds of nozzles:

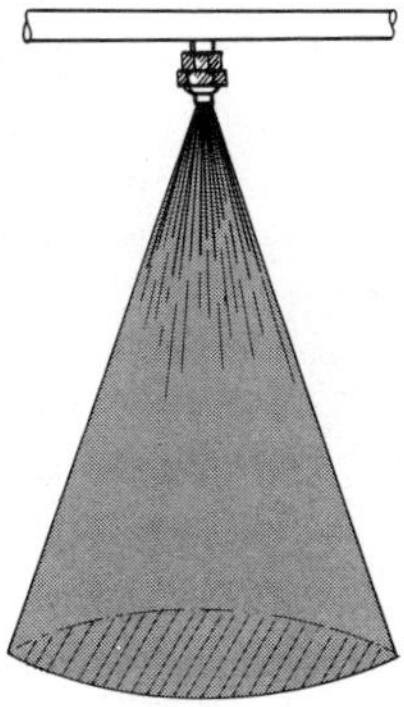

Flat fan nozzle makes a narrow oval pattern with tapered edges. It is used on booms, overlapping with other nozzles by 30-50 percent to form even broadcast pattern.

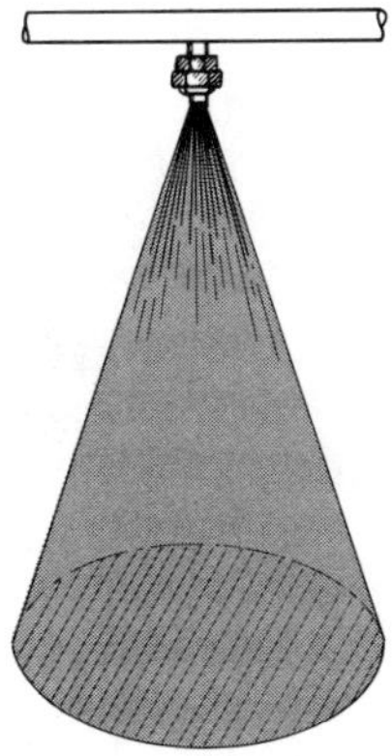

Solid cone nozzles are used for spot treatment.

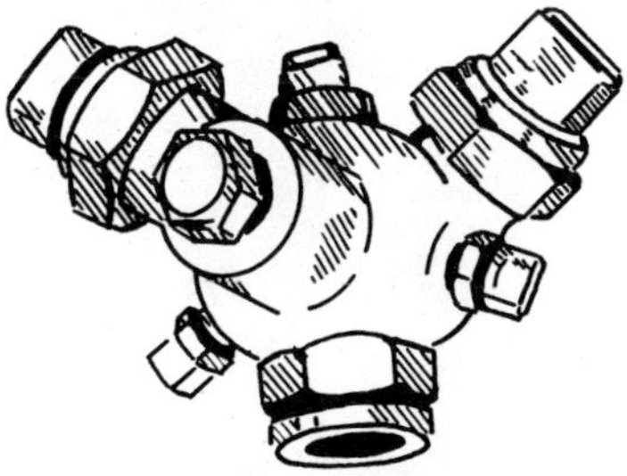

Broadcast nozzles are used where a pattern is to be extended beyond the end of a boom, or to make boomless sprayer.

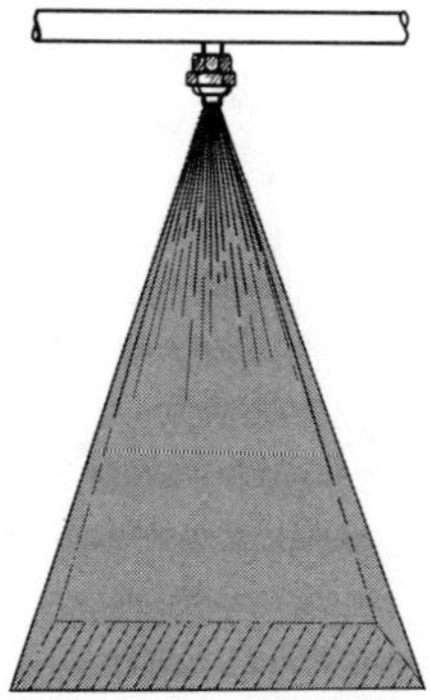

The *even flat fan* nozzle makes a uniform pattern across its width, and is used for band or strip spraying.

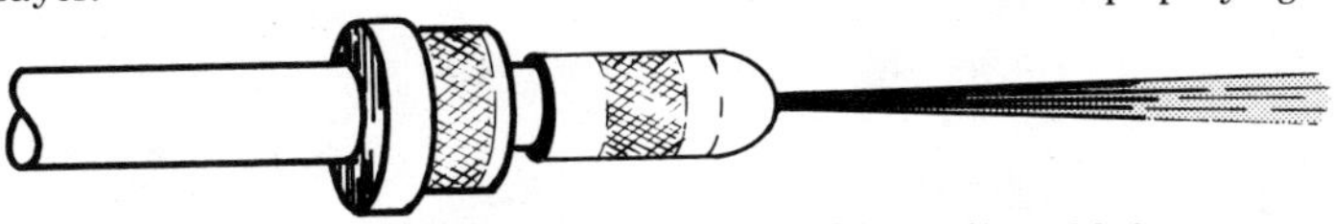

Solid stream nozzles are used on hand sprayers. Often adjustable between stream and solid cone. Used for basal spraying as solid stream, general spraying as solid cone.

The nature of the spray formed by the nozzle is subject to operator control in various ways. To make the spray coarser:

- Reduce pressure by adjusting pressure relief valve
- Increase nozzle orifice diameter
- Increase viscosity of mixture

Aerial equipment is usually equipped with one of the following nozzles:

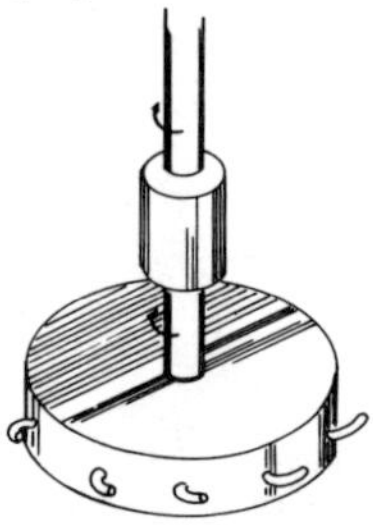

Spray disc used to apply invert emulsions and thickened materials in large drops.

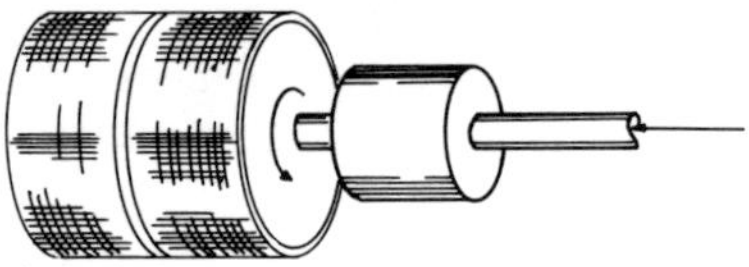

Micronair spin nozzle for applying ultra-low volumes of insecticides in small droplets.

Raindrop® nozzle. For use where large drop is needed for control of drift

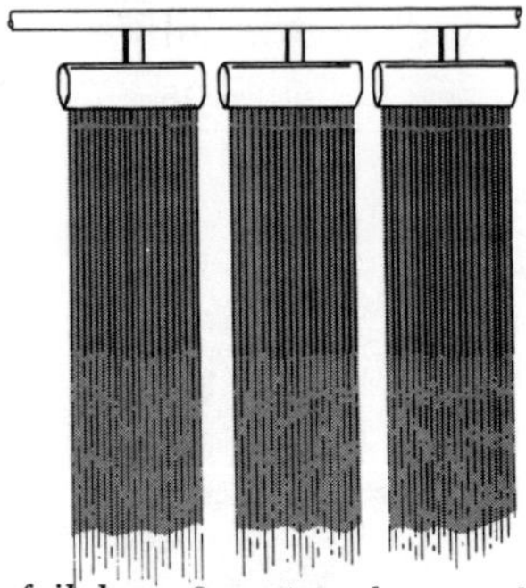

Microfoil boom® system has needle-like orifices projecting into air foil to minimize formation of satellite droplets.

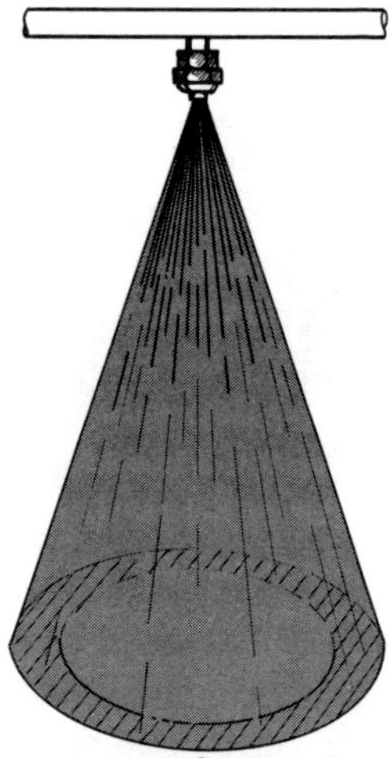

Hollow cone nozzle may have D-2, D-4, D-6, D-8 or D-10 orifices to adjust for very fine to very coarse spray. Number is diameter, in 64ths of an inch

- Remove whirlplate from nozzle, if present
- Direct nozzle to spray directly back along airstream

Drift Properties of a Spray

Sprays used for different purposes must have different properties. These properties determine coverage and effectiveness, and also the tendency to *drift,* or move off target. The most important properties that influence drift are mean droplet size, variability of droplet size and percentage of volume in very small, drift-susceptible droplets.

A very small volume of liquid, as used in ultra-low volume (ULV) applications, must be dispersed in extremely small droplets in order to achieve adequate coverage. Insecticides using such technology typically are released in sprays where most of the volume is in droplets with diameters in the range of 40-160 microns and a volume median diameter (VMD) of about 100 microns. Such a droplet size distribution will provide a pattern of 16-19 droplets per square centimeter (100 to 125 droplets per square inch) in swaths up to 2,000 yards wide if applied in a steady 8-10 mile-per-hour crosswind. This is regarded as a nearly ideal pattern for control of defoliators, which are controlled according to *droplet density* rather than quantity of chemical on target. It has obvious limitations for use on small ownerships.

The fine droplets used for insecticides on very large projects are not suitable for herbicides. Drift can be a major problem with herbicides, and the effect of material on any given leaf is directly related to quantity deposited. Typical herbicide applications of ten gallons per acre are delivered with a VMD of about 350-450 microns. The proportion of droplets below 200 microns is very small, and drift potential is low in winds less than 5 mph and stable air.

Herbicides differ in their coverage requirements. Soil residual chemicals with little foliage uptake, such as atrazine, need spray volume mainly to provide a suitable vehicle for dispersal. The nature of the droplet has little influence on chemical effectiveness apart from the tendency for small droplets to evaporate before reaching the ground, and high volumes provide no particular advantages. Phenoxy herbicides must be uniformly absorbed by foliage, and coverage with at least 13 droplets/cm^2, (80 droplets/in^2) is needed in sufficient volume so that the drops can flow together to form a film on the leaf surface. Large volumes tend to be more effective than low volumes, and 5-10 gallons per acre is usually recommended as a minimum for aerial application.

Virtually all products formulated for application in water contain one or more surfactants. They enhance the absorption process, and may also aid translocation. They can compensate to a degree for marginal coverage.

Herbicides that translocate very readily, such as picloram or glyphosate, can perform well in lower volumes than phenoxys. Three to five gallons per acre appear to provide satisfactory coverage for these herbicides. They are generally applied in 10 gallons per acre, however, to simplify adaptation of equipment calibration.

Minimum volume may be determined by the minimum size droplets that can be tolerated in terms of drift. As a practical matter, most aircraft are calibrated for herbicides (high volume), or for insecticides (low volume). The equipment is not used interchangeably in most forestry applications, although selection of nozzle size and number allows some flexibility.

Spray drops from a given nozzle always vary in size. When a liquid enters the airstream at 50 to 100 miles per hour, the sudden turbulence causes the tearing apart of drops into a variety of small particles. There are always some fine drops, and the distribution of drop sizes includes all sizes from less than 20 microns to roughly 800 microns in a 400-micron VMD spray (a standard coarse herbicide spray) when applied through conventional cone nozzles with D-6 orifices. If nozzles are oriented into the wind, the average drop size decreases, and the percentage of fine drops increases. If nozzles are pointed toward the rear, away from the wind, drop size increases and the percentage of fines decreases. If the nozzle opening is increased and the pressure decreased, the volume per acre will remain the same but droplet size will increase while fine droplets decrease. Thus, nozzle orientation may be combined with pressure and nozzle size to produce a desirable spray. There is no nozzle system that *totally eliminates* fine droplets, and the *danger of drift is always present* where there is substantial air movement. The volume in fine droplets merely determines *how much* will move, and *how far* it will go on a particular wind. Spraying when air is moving *away* from sensitive boundaries will keep drift of spray droplets from trespassing. The movement of fumes from volatile materials long after the application can cause injury to highly sensitive crops, however, and cannot be prevented by application technology.

Fine droplets are significant for several reasons. For insecticides, they provide a large number of toxic spots on foliage without requiring excessive volume or toxicant. They are an advantage in this respect. For herbicides, fine droplets move off target as they do with insecticides, but cause injury to sensitive plants. They also tend to evaporate while falling because of slow descent and large surface-to-volume ratio. The drying of drops less than 150 microns in diameter is an important cause of chemical loss in herbicide applications released more than 50 feet above the target. These losses are important both as an expense to the user and as unwanted drift. Low temperatures (<70°), high humidity (>50% RH), large droplets and stable air reduce losses.

Two technical developments have provided a degree of control over drift without modifying the spray mixture. The Microfoil^{R} boom does not utilize a conventional nozzle, but replaces it with a series of hypodermic needle-like tubes injecting many smooth streams into an air stream conditioned by the air-foil shape of the boom. The lack of wind turbulence is combined with a low-pressure, smooth stream to provide an even breakup into uniform drops, each of which has a satellite drop. The small satellite slows more rapidly than the large drop, and it is absorbed by the following drop when they collide. There is, therefore, very little size variation in the resulting drops, and they are large enough to fall almost directly to the ground even in a substantial crosswind. The Raindrop^{R} nozzle applies some of the same principles to conventional spray booms.

When using sprays consisting principally of large drops, it is necessary to use large volumes of liquid per acre, or to use materials which are readily translocated. Drift control equipment has the major disadvantage of reducing coverage. More chemical, or more carrier, will be needed when coverage is reduced unless the chemical is easily translocated. Very few of the herbicides and virtually none of the insecticides qualifies in this regard, and the substitution of drift-controlling equipment for conventional nozzles is justified *only* where sensitive crops or boundaries are present nearby, where topography forces higher than normal flying (e.g. steep sideslopes, see p. 141) or where insecticides could drift into streams. Drift control equipment is wasteful of chemical and aircraft time under conditions where there is no danger of drift damage. It leads to excessive costs, and either excessive application rates or poor results. This technology is therefore adapted only for special situations, and is inappropriate for general use.

Drift is not exclusively the result of fine droplets being blown off target. Fumes may move off after spraying volatile chemicals. Ester formulations of phenoxy herbicides, picloram and dicamba can evaporate and cause trouble. Low-volatility esters will provide nearly fume-free results if used during cool weather. High humidity does not reduce vaporization of esters, but evaporation of fine droplets containing water is reduced during application. This prevents drops from shrinking to sizes that do not settle readily in air.

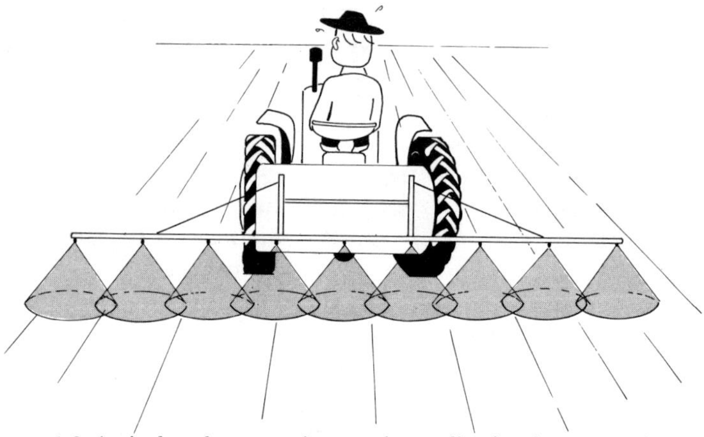

Figure 4-2. Agricultural sprayer gives precise application, but cannot be operated over tall vegetation or on rough terrain.

Application Equipment

Power Sprayers

Applications of pesticides in forestry often require special equipment. Aerial equipment and ground rigs designed for agricultural crops are often poorly designed for the rough terrain and large trees characteristic of forests. Unfortunately, forestry uses have been so minor that development of special equipment has been limited.

Conventional spray booms mounted on tractors are applicable only in open fields or in nurseries. These rigs have evenly spaced nozzles and provide uniform coverage at highly controlled speeds (Fig. 4-2). They will generally provide precision application with negligible problems of drift. Their use on forest land is limited principally by the ability of the tractor to follow a straight line at an even speed on uneven or rough terrain. They have the advantage of great uniformity within a swath, and the disadvantage of leaving skips whenever there is a small guidance error.

Tractor-mounted mist blowers are used extensively on level ground where tractor movement is relatively easy. They are used especially in understory brush control beneath southern pines in well-established stands approaching maturity (Fig. 4-3). Mist blowers, known also as air-blast machines, consist of power-driven fans directed past a nozzle system so as to deliver a relatively fine droplet size spectrum into the air stream. The air stream is projected upwards from the tractor, and is capable

*__Figure 4-3.__ Tractor-mounted mist blowers are useful for understory control in established pine forests.

of controlling brush up to 25 feet tall. The tapered edges of the pattern permit some minor maneuvering along the application path without leaving skips.

Tractor-mounted mist blowers produce heavy operator exposure to the chemicals being used unless special precautions are taken. Broadbrim hats, respirators, goggles, gloves and coveralls are recommended for most chemicals. Some materials registered specifically for this use require such precautions on the label.

Dusters are devices for directing clouds of suspended dust products to an insect problem site. They usually entail an air-channeling device on an aircraft or blower systems that feed dust into the airstream at a calibrated rate. Dusts are losing popularity because of drift and risk of inhalation.

Aircraft

Aircraft apply most of the pesticide volume used in forests. Helicopters are used almost exclusively on herbicide projects, but large fixed-wing airplanes are used extensively for regional insect control projects. Rough terrain places a severe limitation on airplanes, and also certain limitations on helicopters.

Adaptation of airplanes for large-scale insecticide operations is a highly specialized job. Many different kinds of aircraft are in service, ranging from conventional, single engine agricultural rigs to converted bombers and large multi-engine airliners, such as the Lockheed Constellation or DC-7. Each piece of equipment must be adapted to the job requirements. Because the rigging of each type of aircraft for spraying must be adapted to its aerodynamic characteristics, the specifics of adaptation are beyond the scope of this handbook.

Airplanes are effective for applying large amounts of material to very large areas. Depending on the size and type of aircraft and the nature of the job, flying speeds range from about 90 to 150 miles per hour, producing swath widths in still air from 40 to 300 feet. When using ultra-low volumes and small droplet sizes, applied in a crosswind, effective swath width for insecticides may extend 2,000 feet.

Airplanes require an airstrip close to the project. The large aircraft require wider and longer strips than agricultural planes, but don't have to reload as often. Airplanes normally operate 50-150 feet above the canopy when applying insecticides on level terrain. This gives adequate visibility for the pilot and also an opportunity for air movement to make the most out of swath dispersal for maximum coverage. Boundary visibility and aircraft guidance are major problems under the best of circumstances, especially for fixed-wing aircraft operating in long swaths. It is especially important to provide visible markings for swaths and treatment boundaries when using fixed-wing aircraft. Electronic guidance systems are being developed for this purpose, and will permit aircraft to travel on highly specific vectors with great accuracy in the absence of ground markers. In mountainous terrain, flying height is higher and more variable, leading to a more variable deposit.

Helicopters are much more versatile than fixed-wing aircraft for steep terrain or small projects and operating close to the canopy. Helicopters vary in size and equipment, but the most widely used are the small gasoline-powered craft in the 270-340 horsepower range. They normally carry maximum payloads of 60-90 gallons of water, or somewhat more if oil is used as a carrier. Modified with turbine engines, these aircraft can handle heavier loads. Helicopters normally spray at 45-50 miles per hour with an effective swath width of 50-100 feet (Fig. 4-4), and average performance of roughly 60 acres per hour. The larger turbine-powered helicopters have larger payloads and higher flying speeds. They are less maneuverable, however, and so do not cover much more ground when projects are in small patches, as is customary for herbicide application. The large equipment is therefore limited for herbicide application.

The versatility of helicopters is not without limitations. When

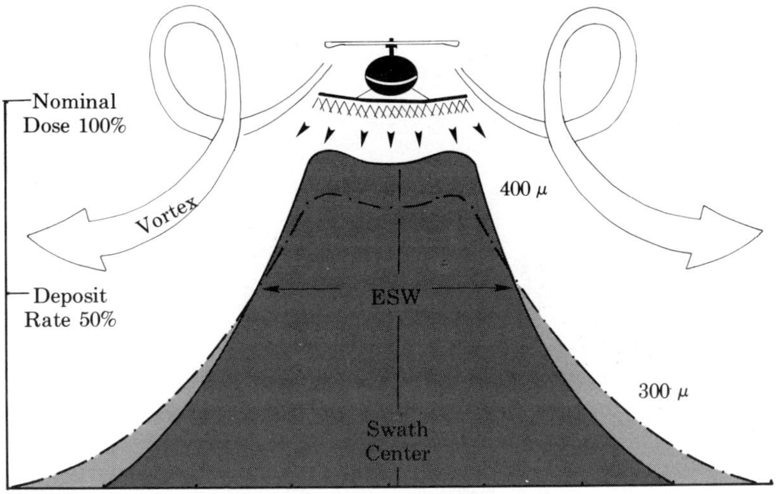

Figure 4-4. Deposit of spray from helicopter applying medium and coarse sprays in still, cool air. Effective swath width (ESW) is distance between swath centers when adjacent swaths overlap sufficiently to form uniform total spray. Shaded areas illustrate actual deposit by coarse (400 u VMD) or medium (300 u VMD) sprays.

applying on steep topography, major difficulties arise in achieving uniform coverage, whether flying the countour or flying up and down hill. Flying the countour of a steep hill leads to a heavier than planned dosage on the uphill edge of the swath and a lighter treatment on the downhill edge. This strip effect means that little of the area is treated correctly.

Operation of a helicopter uphill gives rise to a power effect in the rotor wash. This provides thorough coverage and a somewhat narrower than normal swath. When the aircraft flies downhill, however, the reverse occurs. Thus the orientation of swaths up-down leads to strip patterns of a different kind, which are also undesirable. Helicopters can apply chemicals satisfactorily on steep ground if all spraying swaths are flown either uphill or downhill. This procedure is not as safe as the other approaches, and is far less efficient in the use of the helicopter time because no chemical is released on the return trip. Cost therefore rises, although such increases are a small fraction of the cost of a poor job and a sacrificed plantation.

Helicopters are the only aircraft on which Microfoil[R] and other drift control devices may be mounted satisfactorily, which gives them another advantage in steep topography. Contour flying on steep hillsides, using a drift-control-equipped aircraft

Figure 4-5. Backpack sprayer rigged for broadcast spraying.

substantially above the canopy, provides a reasonably sound deposit pattern. Stripping is avoided because of reduced swath edge vortices that are responsible for the stripping effect of conventional equipment in similar circumstances. High flying and drift control equipment pose difficulties with guidance, however, because there is very little overlap between swaths, and precision of swath estimation by the pilot is reduced.

All aircraft registered for agricultural spraying must be operated by certified applicators. The examination for pilots is the most rigorous of the certification exams, and it is not expected that the resource manager will have to be an expert on either the design or operation of aircraft. The manager's job is to know precisely what he wants the pilot to do, and he should understand the limitations of the equipment in order to avoid false expectation.

Hand-Held Equipment

Hand application equipment for use in forestry is rudimentary. The backpack sprayer, backpack mistblower and tree injectors are all simple devices. They do not operate as rapidly as power equipment, but their versatility and limitations deserve attention, especially for the small woodland owner.

Backpack sprayers have traditionally been used principally for spot spraying brambles, grass around seedlings and for basal bark treatment of hardwoods. Instructions for such uses are normally found in detail on labels that accompany the herbicide

Figure 4-6. Backpack sprayer equipped with even fan nozzle for band applications.

or insecticide. The backpack is not limited to such spot uses, however. It can be adapted for broadcast application by mounting a small boom that projects to the rear, with nozzle spacing to provide suitable overlap between nozzles and between swaths (Fig. 4-5). The spray wand also can be adapted with an even flat fan nozzle tip to function as a very short boom for strip spraying (Fig. 4-6).

The backpack sprayer can also be used for broadcast application by directing the nozzle slightly upward and waving it back and forth rhythmically as the user walks. The swath must be of even width, and walked at an even speed. The nozzle should be oriented upward only enough to insure that droplets are falling nearly straight downward when they hit target vegetation or trees. If the spray hits a tree when close to the nozzle, overdosage and possible damage to the tree may ensue (Fig. 4-7).

The backpack mist blower may be used in the same way as the backpack or garden sprayer, with certain limitations. The mist blower has a smaller capacity tank than the backpack sprayer, usually about three gallons, and it is somewhat heavier than the full backpack sprayer. It projects the pesticide a greater distance than does the sprayer, and produces more fine droplets and better coverage, but more drift. The mist blower should be used in preference to the sprayer when low volume coverage is essential, as for certain insect problems on Christmas trees. Although it can be used for conventional spot and broadcast herbicide application, it has the disadvantages of motor noise and vibration; excessive fine droplets; cost; and excess

Figure 4-7. Use of a backpack sprayer for broadcast application. Uniform pattern requires care and practice.

weight. It cannot be adapted with a boom for uniform broadcast treatments. Applicators of insecticides with mistblowers should wear protective equipment, and avoid inhalation of mist.

Tree injection equipment varies from the hatchet and squeeze bottle, at the simplest, to an automatic hatchet injector, which automatically injects liquid as the instrument hits the tree (Hypo-Hatchet). Regardless of the type of equipment, it operates on the basic principle that injection requires a measured amount of concentrated chemical placed within the inner bark or outer sapwood of the target tree. Dosage level is varied by adjusting the distance between injections.

Hatchets and plastic squeeze bottles have been used successfully on small to large scale operations. This equipment is low in cost, and very simple to use. It typically results in relatively low work speed, but negligible wasted time for maintenance or repair. Worker exposure to chemicals is higher than with other tools because of frequent refilling, and saturation of gloves from leaking bottles, falling down and other inevitable consequences of using make-shift equipment with highly concentrated chemicals.

Tubular injectors increase working speed somewhat and also reduce operator exposure to chemicals. Tubular injectors are best suited for level terrain and large individual stems. They

Figure 4-8 a&b. Tree injection equipment in use. Operator should wear impermeable boots and cloth gloves, any time contact with the chemicals occurs in use, and glasses where spattering may occur. Chaps offer protection from equipment leaks when using hypo-hatchet injectors.

are not well adapted for injection of sprout clumps or struggling through dense brush on steep terrain. Injectors of the hypo-hatchet type typically provide the most production per man-day when operated correctly. The complexity of the mechanism requires some training to minimize downtime. This equipment is well suited to clumps of trees and steep terrain because of the one-hand operation. Proximity of the injection to the operator's face can lead to facial spattering with herbicide. If workers are using organic arsencials in terrain where application often entails injection substantially above waist level, facial cream should be applied to minimize dermal exposure. Equipment and suitable protective clothing are illustrated in Figure 4-8.

All applications of herbicides to stumps or other cut surfaces should be done soon after making the cut. Vessels plug within hours of cutting, so treatment the day after felling or girdling is not effective. When treating stumps with a squeeze bottle, the herbicide should be carried by the saw operator, and applied immediately. Immediate application of the herbicide following the creation of a wound by an injector is an advantage of this method.

Chapter 4 References

Newton, M. and J. A. Norgren. 1977. Silvicultural chemicals and protection of water quality. U.S. Environmental Protection Agency Rept. EPA 91019-77-063.

Potts, S. F. 1958. Concentrated spray equipment, mixtures and application methods. Dorland Brooks, Caldwell, N.J.

Prebble, M. L. (Ed.). 1975. Aerial control of forest insects in Canada. Department of the Environment, Ottawa.

Stewart, R. E. and H. Gratkowski. 1976. Aerial application equipment for herbicidal drift reduction. U.S.D.A. Forest Service, Pacific Northwest Forest and Range Experiment Station, Gen. Tech. Rep. PNW-54.

U.S.D.A. Forest Service. 1978. Vegetation management with herbicides. Final Environmental Statement, Region 6, Portland, Vol. 1.

U.S.D.A. and U.S.E.P.A. 1977. Apply pesticides correctly; a guide for commercial applicators. U.S. Govt. Printing Office No. 0588643.

CHAPTER 5

RISKS AND CARE IN HANDLING PESTICIDES

Herbicides, insecticides and all other pesticides are designed to cause injury to certain living organisms. This is why they are useful, but it is also why organisms other than target species may be injured. The questions of which species are injured, how likely they are to be affected, and to what degree they are affected are related to specific biological activity of the pesticide and the nature of exposure. The physiology of the affected organisms also determines whether recovery will occur and how quickly. Applicators are invariably exposed to the materials they handle. Their safety is determined by proper handling precautions in keeping with the properties of the pesticides, and administration of proper treatment if inadvertently exposed to harmful amounts (Appendix, Table 1).

Biological Properties of Pesticides

Pest control agents can be registered (approved) for use only when effectiveness on target species is demonstrated and when human and environmental health are not jeopardized in normal use. Registered pesticides usually affect specific metabolic functions. Species possessing these specified functions are injured when physiologically exposed to a toxic level of the pesticide. For example, pesticides that inhibit acetylcholinesterase affect only species with central nervous systems; photosynthesis inhibitors affect green plants, etc. Despite their specificity, these chemicals are subject to all the general principles of toxicology, and, like nearly all substances, can cause adverse toxic effects for any organism if exposure is sufficiently high.

Chemicals frequently have more than one potential metabolic effect. A secondary effect may appear in an exposed organism at exposure levels much greater than those used for pesticidal activity on a target species when the non-target species is not affected through the primary mode of action. Thus, herbicides designed to control plants through effects on cell wall growth have the ability to affect animals at much higher dosage levels through unrelated metabolic systems. Secondary effects, where they occur, are of importance in determining acceptable exposure levels for non-target species and operator hazards. An understanding of secondary effects is fundamental to establishing suitable safety margins; this is a major focus of modern toxicology research.

The primary effects of most registered herbicides and insecticides are limited to plant-specific and animal-specific metabolic functions. These effects are studied as part of the testing for labelling. The precise mechanisms by which some chemicals produce secondary effects are poorly understood, so investigators must rely on empirical evidence of toxicity and a substantial margin of safety to compensate for this uncertainty. Ordinarily, a "safe" level for humans is regarded as 1/100 of the highest dose administered to several laboratory animal species with no observable effect in any of them.

Insecticides

These chemicals belong to three major chemical families, the chlorinated hydrocarbon, the organophosphorus and the carbamate insecticides. These materials differ in environmental behavior, and to some degree in their toxic effects. Their toxic properties are listed in Table 5-1, along with reference to their grouping for care in case of accidental poisoning, as indexed in Table 1, Appendix.

The chlorinated hydrocarbon (organochlorine) insecticides, endosulfan and endrin, are low in water-solubility and are persistent in soils and organic matter for more than a year. Endrin is highly toxic, but as it is normally used as a rodent repellant in small quantities, environmental contamination and nontarget species impact are rare in forests. Endosulfan, in contrast, is barely in the highly toxic class but is used at relatively high rates for control of aphids and midges on Christmas trees. In Christmas tree culture, its persistence, toxicity and adherence to surface soils suggest that it be used with extreme care, especially if runoff water enters fish-bearing streams. Lindane, another organochlorine insecticide, is more soluble in water than other members of the group, and it degrades more rapidly. Its use is restricted to applications in oil for bark beetle control, in which the principal concern is with operator exposure. The other organochlorine insecticides are limited to spot treatments with narrowly localized environmental impacts.

The organochlorine insecticides affect the function of the central nervous system, but do not inhibit cholinesterase enzymes. They are taken in by target insects either by contact or through the stomach. Similarly, they may enter the applicator via ingestion or through dermal contact.

Organochlorine compounds are soluble in fat or oil. They tend to remain in fat, and therefore may accumulate to relatively high levels with intermittent exposure. Between exposures, there is a gradual decrease in body load. Most people carry low levels of several of these compounds in their tissues constantly without known harm. Chronic exposure to large amounts will cause illness. Determination of safety margins for these insecticides requires long-term toxicological tests.

The tendency for organochlorine insecticides to accumulate in fat leads to foodchain build-up, known as biomagnification. This is the only class of pesticides where such accumulation is known to occur, and to which long-term environmental problems have been attributed. These problems can occur only because deposits in fat are largely immobile, and various organisms can carry large insecticide residues in fat with negligible consequences to their health. When such species are preyed upon, the predator may be exposed repeatedly to high dietary concentrations. In predatory birds, these high concentrations have led to difficulty with reproduction. Unfortunately, however, such a multitude of fat-soluble pollutants have entered the diets of various raptors that the role of each chemcial as a cause of reproductive difficulty is not clear. There is a major body of scientific opinion, however, that implicates DDT and other materials of this group as the primary cause of problems for predatory birds and aquatic species.

Organochlorine insecticides are the most persistent pesticides. Microbial breakdown degrades most pesticides, but the large chlorine atoms that shield the rest of the molecule (steric hindrance) inhibits microbial access to the internal structure of these molecules. Shielding and lack of chemically reactive groups make these chemicals relatively invulnerable to chemical

Table 5-1. Insecticides registered for one or more uses pertaining to the growing of forests or Christmas trees, their classification, toxicity and treatment proce-

Insecticide	Persistence	Insects Mg/cm²
Acephate (Orthene)	Short	.42b
Aminocarb (Matacil)	Short	
Azinphos methyl (guthion)	Short	
Carbaryl (Sevin)	Short	5-9[bg]
Carbofuran	Short	.01-.02[a,s]
Chloropicrin (fumigant)	Short	
Chlorpyrifos (Dursban)	Short	
Demeton (Systox)	Short	
Diazinon	Med.	
Dichlorobenzene (O & P)	Short	
Dicofol (Kelthane)	Med.	
Dimethoate (Cygon[R])	Short	.065, 1-35[b]
Disulfoton (Di-syston)	Med.	
Endosulfan (Thiodan)	Long	
Endrin	Long	
Ethylene dibromide	Short	
Fenitrothion (Sumithion)	Short	.01[S], .31[b]
Gardona	Short	.03[l] 1.45[b]
Imidan	Short	
Lime sulfur	Med.	
Lindane	Long	.73[w]
Malathion	Short	
Metasystox - R	Short	
Methomyl (Lannate[R])	Short	
Methoxychlor (Marlate[R])	Long	
Methyl bromide (fumigant)	Short	
Mirex		
Naled (Dibrom)	Short	
Pentac	Long	
Phorate (Thimet)	Short	
Phosphamidon (Dimecron)	Short	
Propoxur (Baygon[R])	Med.	
Pyrethrins	Short	
Kotenone	Short	
Toxaphene		
Trichlorfon (Dylox[R])		.05-.255

[1]Groupings for care of poisoning: C = Carbamate; CH = Chlorinated hydrocarbon; L = Lachrymators; OP = Organophosphorus; P = Pyrethrins; R = Rotenone; S = Sulfur. See Appendix Table 1 for management of poisoning cases.

dure in the event of massive exposure. Toxicity to insects given where known in terms of foliage coverage.

Acute LD_{50} Rats Mg/kg	70 kg Human Grams	Poisoning[1] care class	Remarks
866	60	OP	Low toxicity to fish
50	3.5	C	
15	1	OP	
500	35	C	
11	.77	C	Low dermal toxicity
-20 ppm vapor -		L	Highly toxic; fumes
163	11.4	OP	
4-12	.28	OP	Systemic, highly toxic
466	32	OP	Toxic to birds
800+	56+	CH	Fumigant, herbicidal
809	56	CH	
215	15	OP	Weak carcinogen
12.5	.87	OP	High dermal toxicity
30-110	1.8	CH	
7-43	.49	CH	High dermal toxicity
146	10.2	CH	Vapor toxic @ 200 ppm
500	35	OP	Low toxicity to fish
4,000	280	OP	<.1 ppm TCDD
216	15	OP	
- no data -		S	Caustic, phytotoxic
88	6	CH	Low dermal toxicity
1,375	96	OP	Low dermal toxicity
105	7	OP	
17	1.2	C	
6,000	420	CH	
- 200 ppm vapor -		L	Fumigant
306	21	CH	Chronic toxicant
430	30.1	OP	
3,160	221	CH	
2-4	.14+	OP	Highly toxic
27	2	OP	
128	9.0	C	
1,500	105	P	
50-500	3.5+	R	Toxic to fish
69	4.8	CH	
500	35	OP	

a - aphids
b - budworm
g - Gypsy moth
w - white pine weevil
s - sawflies

reactions, as well. The combination of persistence and tendency to accumulate in fat during prolonged low-level exposure are responsible for the high potential for biological damage to non-target organisms when these compounds are released in large quantities over large areas.

Organochlorine insecticides tend to be highly toxic to fish, and special care should be taken to keep water contamination at extremely low levels.

Carbamate insecticides are substantially less persistent than organochlorine compounds, and are low in fat affinity. They are in the same range of toxicity as organochlorine materials, but are not known to accumulate. The chief hazard is to applicators and non-target species contacted inadvertently during application.

Carbaryl is the only carbamate insecticide presently used on a broad scale in forests. It is low in mammalian toxicity, low in persistence and moderately high in toxicity to fish. It is very damaging to aquatic insects. It does not biomagnify, however, and even those forest fauna affected by sublethal levels soon recover from injury. Damage either occurs quickly, or not at all.

Carbamate insecticides are inhibitors of cholinesterase, one of the enzymes involved in transmission of nerve impulses. A dose substantially less than the median lethal dose (LD50) will produce symptoms in man if given quickly. Care in handling is imperative.

Many of the insecticides used in forests and on ornamentals are organophosphates. As a group, these are very low in persistence. They are also cholinesterase inhibitors, and affect acetylcholinesterase function in the nerve junctions with muscles. They are absorbed effectively through skin and stomach, either of target species or the handler. There is a wide range of toxicity among these compounds, and they include the most and the least toxic chemicals used in forest insect control. Acute toxicity data are appropriate for estimation of safety for applicators or non-target species.

Organophosphate compounds are generally oil soluble, but they are not persistent enough to collect in fat. Therefore, they do not biomagnify.

Organophosphates are generally toxic to insects in very small quantities. Aquatic insects are no exception, and contamination of waterways containing fish populations may lead to immediate decreases in stream insects at concentrations that do not appear to harm fish directly. Such effects are generally transient, however, because stream insects are normally present in all life stages, of which one or more may be resistant to intoxication. Stream recovery after accidental contamination is usually rapid, barring severe fish kills which are rare.

Applicators should use great caution when handling all insecticides, especially the highly toxic products. They are active on metabolic systems that are important parts of human physiology, and may be absorbed through the skin, inhaled in dusts, or taken by mouth if crews are careless when washing.

Insecticides account for a great many human poisonings each year. Most occur among children who drink from unlabeled bottles or among applicators who have become careless. Some of the most toxic organophosphates are absorbed through the skin with harmful effects, even in the tiny quantities encountered by farm workers entering fields immediately after application. Be careful! Table 5-1 and Appendix Table 1 describe toxic properties of the insecticides and how to treat persons who have had heavy exposure.

As general principles:

- Wear gloves, coveralls, hat and respirator when handling concentrated insecticides with LD50's less than 100 mg/kg.
- *Always read the label,* in its entirety, before mixing. The label outlines proper precautions to be taken in mixing and applying. This is for the protection of applicator, crop and non-target organisms.
- Never overdose.
- Always clean up promptly if exposed.

Herbicides

There are many classes of herbicides. Each class has certain general effects on plants and also a generic behavior pattern in the environment. Their grouping by activity in target plants is a reasonable indicator of activity in sensitive non-target plants. Major effects on non-target animals, fungi and other organisms are generally limited to massive spills or grossly improper use. Unlike field-use rates of insecticides, which may cause direct injury to non-target animals, field-use rates of herbicides are not normally high enough to produce effects on humans or animals through primary or secondary intoxication. In the event of accidental massive exposure, however, Table 5-2 and Appendix Table 1 list toxic properties and proper management of poisoning cases.

Table 5-2. Herbicides registered for one or more uses pertaining to the growing of forests, nurseries or Christmas tree, their persistence, toxicity, and grouping

Herbicide	Persistence[1]	Rats-mg/kg
Amitrole	Short-med.	1,800
Ammonium sulfamate	Short	3,900
Asulam	Short	> 5,000
Atrazine	Med.	3,080
Bifenox	Med.	> 6,400
Cacodylic acid	Short	1,350
DCPA	Med.	3,000
Dalapon – Na	Short	7,570
Dicamba	Med.	800
Dichloroprop	Short	2,900
Dinoseb	Short	50-60
Diphenamide	Med.	1,000
Fosamine	Short	> 24,000
Glyphosate	Short	4,320
Hexazinone	Med.-long	1,690
Mineral spirits	Short	
MSMA	Short	1,800
Napropamide	Med.	> 5,000
Paraquat	Short	150
Picloram	Long	8,200
Pronamide	Med.	8,350
Silvex	Short-med.	650
Simazine	Long	> 5,000
Triclopyr	Med.	710
2,4-D	Short	300-800
2,4,5-T	Short-med.	300

[1]Persistence expressed as half-life. Short ≤ month. Med. 1-6 months. Long ≥ .

The overwhelming influence of herbicides on non-target species is through habitat change. Food availability, cover, species composition and eventual domination of a site by conifers have profound and long-term effects far more important than any known direct toxic effects. These changes are normally a part of the management system, however, and are only incidental to the use of herbicides.

The one herbicide capable of producing direct intoxication to forest fauna is dinoseb. It is of the dinitrophenol group, and is a general biocide. Its metabolic function is to uncouple oxidative phosphorylation, a respiratory process that operates in all living organisms. Its access to metabolic systems is limited only by the amounts deposited, and ability to penetrate into living cells. It is active as an herbicide, fungicide, insecticide and perhaps rodenticide. It is, therefore, used only in preparation for

for management of intoxication in the event of massive exposure.

Acute LD_{50} 70 kg human (grams)	Poisoning care[2]	Remarks
126	General	Non-cumulative
273	General	"
>350	Carbamates	"
215	Triazine	"
>448	General	"
94	Arsenic	"
210	General	"
530	General	"
56	General	"
203	Phenoxy	"
3.5	Dinitrophenols	"
70	General	"
1,680	General	"
302	General	"
118	Triazine	"
30 ml	Hydrocarbons	"
126	Arsenic	"
>350	General	"
10.5	General	"
574	General	"
584	General	"
45	Phenoxy	"
>350	Triazine	"
50	General	"
21	Phenoxy	"
21	Phenoxy	"

[2]Grouping for care of poisoning may be found in Appendix Table 1.

burning. Since it has a short environmental life, and is not translocated by plants, its activity is not systemic, and long-term injury to any organisms is incidental to that of the general disturbance, of which fire is the most important component.

Dinoseb has a low acute oral LD_{50}, and may be absorbed by the skin. Applicators should handle it with due caution. Effects on animals or humans occur quickly if dosage is high enough to produce any symptoms. Recovery is rapid.

Organic arsenicals have one special property that demands care in handling: they are arsenic compounds, and can be converted to the highly toxic trivalent arsenic compounds of mystery story fame. These transformations require a certain environment, such as the presence of a zinc catalyst and acid medium. The organic arsenicals, MSMA and cacodylic acid, are not very toxic, and occur naturally. But they are converted either by catalyzed physical reactions, or by enzyme systems, to highly toxic and gaseous arsines. Under normal circumstances, these remain in small quantities, but the user should be alert to strong smells of garlic when working in an enclosed place.

Organic arsenicals may be absorbed through the skin. Users will often smell garlicky odors on their own breath, an indication of some degree of exposure. This is not a red flag warning, however, as it is difficult to absorb a toxic level through skin. As a prudent measure, it is a good idea to wear clean gloves when handling the concentrates, and to wash well after use. Facial creams reduce absorption. Arsenicals are eliminated quickly by the kidneys.

Paraquat is not used in forestry-related operations except in occasional nursery procedures. It does have a peculiar toxicological property that warrants note, however. Paraquat itself is of moderate acute toxicity. It is possible to be exposed to a lethal dose, however, without being aware of incipient intoxication. Absorption of a toxic dose is possible through skin or stomach over a protracted exposure period and may lead to development of a fibrous thickening of the lung tissues. This condition is irreversible, and in heavily dosed persons leads to pulmonary edema and pneumonia. The manufacturer is making progress toward a management procedure for paraquat intoxication. If heavily exposed, a person should immediately establish communication with a poison control center or the manufacturer.

The phenoxy herbicides have been the subject of much publicity in recent years, especially in relation to the cause of birth defects and spontaneous abortion. Actually, the phenoxys are moderate to low in toxicity to mammals, birds, insects and many species of plants. Unintended effects are limited to those arising from inadvertent contact with desirable broadleaved crops and ornamental trees. Documented illnesses of humans have been rare, and have been confined to persons massively exposed to 2,4-D. Much is known about these chemicals.

The phenoxy acids are low in toxicity to fish. Certain ester formulations are much more toxic than the parent acids. As a general rule of thumb, esters with a butyl group (i.e., butyl, propylene glycol butyl ether, butoxy ethanol, etc.) are damaging to fish when concentrations of one or more parts per million occur in water for 24 hours or longer. These concentrations and durations of peaks are virtually never reached in forest applications, even with accidental application directly to streams.

Thus, damage to fisheries is highly unlikely. Fish injury is virtually impossible through field exposure to esters that lack a butyl moiety, or with any of the amine salts. Silvex and 2,4-D are both registered for aquatic weed control, and are not known to injure fish when used deliberately in concentrations of several parts per million, the normal aquatic use rate which is about a hundred to a thousand times higher than typical concentration maxima in forest streams.

Suspected human exposure to phenoxys can be confirmed readily. These herbicides are eliminated as unmetabolized acids in urine, with a biological half life of 23 hours after oral exposure and 24-48 hours after dermal exposure. Analysis of urine samples taken 24-48 hours after exposure will give an excellent indication of degree of exposure. A dose high enough to produce symptoms (more than 2-4 grams) will show up as several hundreds of parts per million in urine. Concentrations of a few parts per million have little, if any, toxicological meaning.

Silvex and 2,4,5-T contain traces of a contaminant, 2,3,7,8-tetrachlorodibenzo-p-dioxin (TCDD). This contaminant is extremely toxic. It is present in such small concentrations, however, that it does not affect the toxicity or toxic hazard of the commercial products in which it is found. Products containing TCDD now contain approximately 1/1,000 of the concentration that was present before 1964. The early products contained enough TCDD to produce a minor effect on the toxicity of 2,4,5-T, in that high dosages of contaminated 2,4,5-T caused somewhat more birth defects in laboratory mice than high dosages of herbicide containing no TCDD. In any event, birth defects were demonstrated only at dosages far beyond those to which applicators would be exposed without actually drinking the chemical. There is considerable indirect evidence that humans are more tolerant of TCDD than are many other mammals.

The other herbicides encountered in forestry operations are low in toxicity or are non-toxic.

Potential for Hazard

There is some risk in every human endeavor. The degree of risk varies, depending on the activity and the organisms involved. This principle extends to the use of pesticides, where risk may be assessed for the applicator, nearby persons, non-target species, and nearby crops. Risk to the applicator is direct, and is conditioned by the properties of the pesticide and the degree of personal exposure (actual intake). The risk to nearby persons is determined by the degree of inadvertent contact, the properties of the pesticide, and the level of knowledge about actual exposure and how to handle intoxication. Risks to non-target species and crops are determined by properties of the pesticide, level of dosage and application precision.

As a general problem, there are both real and imaginary risks involved with pesticides. Pesticides are potential poisons to various kinds of organisms, but all are harmless at some low level of exposure. Quantities below the threshold of activity merely serve notice of how much of a safety margin exists. Neither health hazards nor environmental damage will occur when chemicals are present below the threshold of toxicity. While it is a serious mistake to overlook a potentially serious exposure to a pesticide, it is almost as serious to attribute some symptoms to a pesticide without solid evidence. A correct solution to a problem can only follow a correct diagnosis. The assumption of toxic effects being caused by a detectable, but sub-threshold, amount of a pesticide will lead to an incorrect diagnosis and inability to deal with the actual cause of illness.

A possible exception to the threshold concept exists with chemical carcinogens. There are several mechanisms involved in the induction of cancer by chemicals, and a scientific concensus has not been reached as to whether some or all chemical carcinogens obey the threshold dosage concept. Therefore, exposure to these chemicals should be kept to a minimum.

Human Exposure

Most pesticide poisonings occur as the direct result of accidental ingestion. Children have a tendency to drink out of beverage bottles and miscellaneous containers, regardless of what is in them. A great many are poisoned by drinking pesticides placed in unlabeled containers and stored within reach. A large percentage of such incidents are fatal because of the massive dosages possible and small size of the victim.

The most common type of human exposure to pesticides is the result of applicator contact, principally dermal and inhalation. Most of these incidents do not cause intoxication, although contact inevitably leads to some degree of physiological exposure. Exposure to pesticides can come through dermal contact alone. Pesticides differ in their ability to penetrate skin, but all are absorbed reasonably efficiently through the stomach. Intoxication through dermal exposure is largely limited to contact insecticides. When a pesticide has been absorbed through the skin, it will be introduced into the metabolic system more slowly than when ingested. This gives the body more time to eliminate the chemical or break it down. But it also renders the chemical inaccessible; once absorbed, the dosage must be dealt with.

Significant dermal exposure in forest operations is almost exclusively confined to applicators. Persons handling technical concentrates in the process of mixing, and individuals using backpack spray equipment, are the most heavily exposed. When using insecticides absorbed readily through the skin, workers should wear protective clothing, such as coveralls, gloves and hat. A supply of soap and water should be kept on hand for washing up in the event of spillage or accidental contact. The most important principle for reducing danger from skin contact is to remember that absorption occurs over a period of time. Rapid removal will minimize absorption.

Among the insecticides, organophosphorus and carbamate materials can be assumed to have the ability to penetrate skin enough to cause intoxication. The most toxic of the organochlorine materials should also be handled with extreme care. Among the herbicides, dinoseb and paraquat should be handled with the same precautions as the moderately toxic insecticides. The other herbicides have very limited ability to penetrate skin. Phenoxys, for example, appear to require repeated immersion of large body parts in concentrated materials to effect sufficient absorption to produce symptoms. Symptoms have been reported for 2,4-D in a limited number of agricultural applicators. Clinical reports of intoxication from forestry herbicides other than 2,4-D, paraquat and dinoseb are lacking, and intoxication by those materials has not occurred while being used in forests. It is particularly noteworthy that many of these materials, particularly 2,4,5-T, have been in use since 1948 in situations where applicators have been heavily exposed, and neither short nor long-term health problems have appeared in the clinical literature.

Persons other than applicators present a special problem associated with involuntary exposure. Some individuals feel that any exposure to a pesticide is to be avoided, and that any exposure is sufficient to bring on malaise. A variety of ills may be attributed to pesticides when nearby applications are in progress. This leads to the problem of ascertaining whether the individuals who are complaining have actually been exposed, and whether some problem unrelated to the pesticide is causing them to suffer.

Clinical treatment of any illness depends on correct diagnosis. It is unwise to administer treatment for poisoning unless it can be demonstrated that some exposure did occur, and that such exposure was at levels consistent with the onset of symptoms. But it is also not possible to obtain lab analyses for body tissues or fluids soon enough to pinpoint the source of trouble. Under the circumstances, a system of managing cases of suspected poisoning has been devised, and recommendations are available from poison control centers throughout the United States. A comprehensive summary of procedures for management of poisoning has been published by Micromedex, Inc., of Denver, Colorado, under the direction of B. H. Rumack. It is known as the Poisindex[R], and is updated every three months. A condensed summary of the management regimes indicated for pesticides used in forestry as of January, 1979, has been reproduced in Appendix Table 1, with the permission of the publishers. This summary is condensed, however, and should be consulted only for interim guidance in handling poisoning cases until contact can be made with a poison control center and appropriate medical personnel. With these guidelines, any competent physician can administer adequate care until comprehensive treatment can be provided by a poison control specialist.

The best precaution against poisoning is the avoidance of heavy exposure. For pesticides with a LD_{50} less than 50 mg/kg, and for endosulfan, toxaphene, dinoseb and paraquat:

- Wear protective clothing when handling, especially gloves, hat and respirator while spraying. Be certain to wear hat and respirator when handling toxic dusts.
- Wash immediately after contact has occurred.
- Change clothes immediately after work.
- Launder contaminated clothing separately.
- If exposed heavily, get medical help immediately.
- Use interim management guidelines in Appendix Table 1 until comprehensive medical attention can be given.

For less toxic substances:

- Wear protective clothing when handling technical concentrates. Gloves are especially important. Wear respirator when handling dusts.
- Wear gloves, hat and long sleeves when spraying.

- Wash immediately in the event of a spill in which major areas of skin are soaked.
- Wash before eating.

Non-target Species

Species other than the intended targets are exposed, particularly when pesticides are applied broadcast. Effects on organisms within the treatment zone are discussed earlier in chapters on insect and vegetation management. Off-site damage, however, may be of considerable importance, and pesticide users have a moral and legal obligation to avoid trespass.

Off-site movement of a pesticide may occur through three principal routes. The first, and most important, is movement of fine droplets downwind. This problem is especially prevalent with herbicides, because the symptoms of injury to crops are readily visible and may have considerable economic impact on the grower. Phenoxy herbicides and other growth regulators are damaging through drift of this kind, because low dosages will cause formative effects on sensitive crops such as beans, tomatoes, cotton, potatoes, and grapes. A dosage one to ten percent of a nominal rate of application will have no apparent effect on the forest target area, but it can devastate one of these sensitive crops, either in commercial production or in a home garden. Although there is a legal machinery to compensate for lost crops, it is far better to avoid the damage in the first place by careful application.

Off-site movement of droplets can be minimized in several ways. To begin with, droplets should be kept as large as is consistent with the pest problem. With phenoxy herbicides, maintaining a minimum of droplets in the sub-250 micron droplet size range will minimize drift damage. Application near sensitive areas can often be scheduled when crops are not in the ground, either in spring or fall, without jeopardizing effectiveness seriously. When it is necessary to control vegetation near sensitive crops, special care should be taken to treat when air movement is noticeably away from the sensitive area.

It is good practice to discuss broadcast treatments with any neighbors who could conceivably be directly affected. These should be informal communications, if possible. Posting of property or public notices may be required by local regulations.

The second opportunity for off-site movement is the drift of fumes from volatile pesticides. Phenoxy herbicides, dicamba and picloram may be formulated as esters. All esters have some volatility, and may be subject to vaporization during warm weather after treatment. Movement of vapors off the treatment area can lead to drift problems similar to those encountered with fine droplets. But such drift can occur during warm weather after application is complete. These problems have been largely solved in forestry with the elimination of highly volatile esters for forestry use. It is noteworthy, however, that there has been an increase in grape culture in forested areas in recent years, and grapes are extremely sensitive to phenoxys. Fumes from even low-volatile esters can injure grapes at some distance from the target during hot weather.

Off-site movement of insecticides will not damage crops directly. They can result in the deposit of residues on crops that are not permitted in food, however. This is a legal matter rather than a toxicological problem, because crops may not be sold in interstate commerce if they have excessive pesticide residues on them, or residues for which no tolerance has been established. In recent years, organic farming has become popular in forested areas. Residues of pesticides are not normally accepted on organic foods. Because there can be no proof that aerial application will deposit no residues whatever on such crops, the existence of organic food growing operations near forests may constitute a land use conflict that must be anticipated in silvicultural planning.

The third mechanism of off-site movement of pesticides is in water. If and when pesticides reach water, they do so almost exclusively through direct application to open water surfaces. Extensive research has shown that neither insecticides nor herbicides are likely to move in soil or runoff after normal applications. Exceptions to this have occurred when 1) applications have been made to bare soil, and have been followed by heavy rains and runoff of silt, and 2) when applications have been followed by heavy rains before the chemical has had opportunity to dry on foliage and ground cover.

Trace residues frequently reach streams as the result of drift and inadvertent overflights. After deposition in water, the chemical moves downstream and is degraded or adsorbed. Characteristics of the stream influence the rate of elimination and the behavior of the contamination "spike" downstream (Table 5-3).

Chemicals are not completely restricted to the immediate vicinity of an aerial application swath; "Tails" of a swath may

Table 5-3. Stream factors affecting the duration and concentration resulting from a pesticide contamination.

Factors Affecting Duration	Factors Affecting Concentration
Ratio of pool to riffle volume	Stream velocity in the area (rapid flow decreases concentration if application period is prolonged)
Stream bottom roughness (rough decreases)	Depth of input water where applied (decreases in deep water)
Temperature (high decreases)	Riparian cover in project (interception decreases)
Adsorptive capacity of bottom (high exchange capacity decreases)	
Stream velocity (high decreases)	
Organic matter in stream (high decreases)	
Riparian cover below project (open, shallow streams subject to photodegradation)	

be detected some distance from the center line, owing to the relatively long period fine droplets are suspended in air. Applications can be conducted, however, so as to keep water contamination well below concentrations capable of causing harm to water users or aquatic organisms. The guidelines for application of forest chemicals near water have been prepared by the U.S. Environmental Protection Agency, and are available in Report EPA 910/9-77-036, *Silvicultural Chemicals and Protection of Water Quality,* available from the National Technical Information Service, Springfield, Virginia, 22161. Table 5-4 is reproduced from this manual, and lists maximum permissible tolerances recommended for streams of various sizes and uses. The chemicals listed are the forest pesticides that were registered for aerial application in 1977. Individual states may have forest practices acts that specifically prescribe the manner in which pesticides may be applied near water. Commercial applicators are required by law to be familiar with the local laws regarding water.

There are several sources of serious water contamination from pesticides:

- Aerial application of insecticides, particularly organochlorine materials, near water.
- Filling batch trucks with equipment that permits mixed material to flow back into the stream or pond.
- Container disposal.

Table 5-4. Recommended concentration maxima for silvicultural chemicals by stream class and user group. Potable waters include safety factors for wildlife

Class	Chemical	Most Sensitive Test Species Affected	Test Basis & Concentration
Fertilizer	Nitrate	Man	No effect, 10 mg/l N
	Phosphate	Algae	Growth response var.
Herbicide	Amitrole	Daphnia	LC_{50} 48 hr, 3 mg/l
	Ammonium ethyl carbamoyl phosphonate	Bluegill	LC_{50} 48 hr, 670 mg/l
	Arsenicals (organic)	Man	No effect, 0.12 mg/l
	Dalapon	Daphnia	LC_{50} 48 hr, 11.0 mg/l
	Dicamba	Bluegill	LC_{50} 96 hr, 23. mg/l
	Dinoseb	----------------inadequate data------	
	Picloram	Bass	LC_{50} 48 hr, 19.7 mg/l
	Silvex[1]	Chinook salmon	LC_{50} 48 hr, 1.2 mg/l
	Triazines	Daphnia	LC_{50} 48 hr, 1.0 mg/l
	2,4-D[1]	Bluegill	LC_{50} 48 hr, 1.0 mg/l
	2,4,5-T[1]	Bluegill	LC_{50} 48 hr, 1.4 mg/l
Herbicide	TCDD	Coho salmon	No effect 96 hr, .000000056 mg/l
Insecticide	Carbaryl	Stonefly	LC_{50} 48 hr, .0048 mg/l
	Diazinon	Daphnia	LC_{50} 49 hr, .0009 mg/l
	Disulfoton	Stonefly	LC_{50} 48 hr, .005 mg/l
	Endosulfan	Rainbow trout	LC_{50} 96 hr, .0003 mg/l
	Endrin	Coho salmon	LC_{50} 96 hr, .0005 mg/l
	Fenitrothion	Atlantic salmon	Behavior test 1 mg/l
	Guthion	Stonefly	LC_{50} 96 hr, .0015 mg/l
	Lindane	Brown trout	LC_{50} 48 hr, .002 mg/l
	Malathion	Daphnia	LC_{50} 96 hr, .0018 mg/l
	Phosphamidon	Daphnia	LC_{50} 48 hr, .0088 mg/l
	Trichlorfon	Stonefly	LC_{50} 96 hr, .016 mg/l

*As listed by the Environmental Protection Agency's Quality Criteria for Water, 1975.

[1] The phenoxy herbicides may occur in water as esters or other forms. The given criteria for potable water may be increased by a factor of 10 for forms other than esters. Criteria for irrigation use are for total phenoxy herbicide.

The above types of contamination with insecticides have caused extensive fish kills and temporary loss of aquatic food organisms. In minor degree, they have caused loss of stream productivity and contributed residues to fish that have taken some time to dissipate. These types of contamination are avoidable and do not contribute in any way to good resource husbandry.

and aquatic organisms as well as humans.

Criteria, PPM 24 hr. Mean Stream Class & User					
≤ 10 cfs		10 cfs-Navigable		Navigable	
Potable	Irrig.	Potable	Irrig.	Potable	Irrig.
10*	10*	10*	10*	10*	10*
--------inadequate basis for recommendation--------					
.15	0.1	.03	.01	.015	.01
5	5	1	1	0.5	0.5
.1	.1*	.05*	.1*	.05*	.1*
.5	.1	.1	.02	.10	.02
.2	.004	.05	.002	.01	.001
inadequate data -------------------					
.5	.001	.05	.0005	.005	.0001
.06	.02	.03	.02	.01*	.01*
.05	.05	.03	.03	.01	.01
.05	.05	.05	.02	.01	.005
.06	.02	.03	.02	.01	.01
-------------.000000006 for all water-------------					
.001	.001	.0005	.0005	.0002	.0002
.0001	.0001	.00005	.00005	.00001	.00001
.001	.001	.00025	.00025	.00024	.00025
.00003	.00003	.00001	.00001	.000003	.000003
.00005	.00005	.00001*	.00001*	.000005	.000005
.025	.025	.01	.01	.005	.005
.0003	.0003	.0002	.0002	.00007	.00007
.0001	.0001	.00005	.00005	.00001*	.00001*
.0005	.0005	.0002*	.0002*	.0001	.0001
.0005	.0005	.0005	.0005	.0002	.0002
.002	.002	.0005	.0005	.00005	.00005

Low-level contamination of water with herbicides and certain of the "selective" insecticides does not cause harm. It may, however, be in violation of the law. The maximum contamination levels published by the Environmental Protection Agency provide a substantial margin of safety, and Table 5-5 lists the recommended application procedures that will maintain water quality within the tolerances recommended.

Table 5-5. Guidelines for Applying chemicals by aircraft, and water monitoring

Practice	Chemical Used	Minimum Distance Between Nearest Water and Center Line of Nearest Swath
Fertilization	Urea	¾ of an effective swath width (ESW).
	Phosphorus	¾ ESW* Exceptions: upstream from lake or impoundment.
Forest Site Preparation	Amitrole	½ ESW* Exceptions: within a mile of potable users, 50-foot buffer.
	Ammonium ethyl Carbomyl phosphonate	½ ESW*
	Atrazine	½ ESW* Exceptions: scarified areas, 50 feet.
	Dalapon	½ ESW
	Phenoxys	½ ESW
	Endosulfan	4 ESW* or 300 feet, whichever is greater.
	Fenitrothion	1 ESW*
	Guthion	3 ESW* or 200 feet, whichever is greater.
	Malathion	"
	Phosphamidon	1 ESW* or 100 feet, whichever is greater.
	Trichlorfon	"
Rodent Control (Seeding)		
Chemical	Endrin	¾ ESW

*ESW, (effective swath width) is the maximum distance between adjacent swath centers without a strip of discernibly low application deposit between swaths. Designation of "None" or ½ ESW under *Buffer Strip Width* implies only that buffer strip width is at the discretion of the operator, and that direct impact on water quality is not at issue. Even without a buffer strip, the aircraft should *never* be operated within a half-ESW of streams that are likely to have fish in silvicultural practices.

Treatment of Buffer	Suggested Location and Frequency of Water Sampling
Apply by ground rig.	Composite, Day 1, at potable user site, if within 1 mile downstream from project.
Apply by ground rig.	None
a) Apply by ground rig. b) Apply substitute chemical. c) Plant buffer zone with tolerant tree species.	Composite, Days 1 & 2 at potable user site if within 1 mile downstream.
Can be treated.	Composite, Day 1 at potable user site if within 1 mile of project downstream.
Do not disturb soil within buffer zone.	None
Do not disturb soil within 50 feet of creek.	None
Can be treated.	Composite, Day 1, at intake if potable user within 1 mile of project downstream.
May treat with carbaryl diazinon, fenitrothion or phosphamidon to 1 ESW from water.	Sample as with organophosphorus insecticides, but sample also after each heavy rain for next month.
May use biological agent.	Same as carbaryl.
"	"
"	"
"	"
"	"
Can be treated by hand.	None

in them at time of chemical application. For those insecticides requiring one or more effective swath widths, the proposed buffers are for helicopters with droplet size of 200 u MMD. If droplets are smaller or large fixed-wing aircraft are used, buffers should be 200 feet plus the given swath numbers. Helicopters may be used in conjunction with large aircraft.

Chapter 5 References

Brown, A. W. A. 1978. Ecology of pesticides. John Wiley, New York, 525 p.

Dost, F. 1977. Toxicology of phenoxy herbicides and hazard assessment of their use in reforestation. U.S.D.A. Forest Service, San Francisco.

Kearney, P. C. and D. D. Kaufman (Eds.) 1969. Degradation of herbicides. Marcel Decker, New York.

National Research Council of Canada. 1978. Phenoxy herbicides – their effects on environmental quality with accompanying criteria for TCDD. NRCC #16075, 443 p.

Newton, M. and J. A. Norgren. 1977. Silvicultural chemicals and protection of water quality. U.S.E.P.A. Rept. EPA 910/9-77-036, 224 p.

Newton, M. and L. A. Norris. 1976. Evaluating short and long-term effects of herbicides on non-target forest and range biota. Down to Earth 32(3):18-26.

Norris, L. A. et al. 1979. Biological and ecologic assessment of 2,4,5-T. A report of USDA, EPA and Coop. States Exp. Stations, USDA, Washington, D.C.

Shepard, T. W. 1973. Handbook of teratogenic substances. Johns Hopkins University Press.

U.S.D.A. Forest Service. 1978. Vegetation management with herbicides. Final Environmental Statement, Region 6, Portland, Vol. 1.

APPENDIX

Management of Poisoning Cases

People invariably are exposed to all substances in general use, some of which have appreciable toxicity. Pesticides are safe to handle as directed, but accidental exposure can exceed safe limits. The following guide to management of poisoned individuals, (Appendix Table 1) will provide an interim procedure until qualified medical help is obtained. It is always wise to seek medical help from a specialist if exposure is more than 10 percent of the median lethal dose as listed in Tables 5-2 and 5-3, or if children take on an unknown dose.

For most poisons, there is no specific antidote, but merely a treatment procedure. In all cases, the victim should be treated in such a way as to make the normal defense systems operate most effectively. Treat the patient, not the poison!

Keep all toxic materials out of the reach of children.

The National Pesticides Telecommunications Network provides 24-hour consultation for management of pesticide poisoning.

Appendix Table 1. Principal components of managing intoxication cases resulting from exposure to forest pesticides. Condensed from Rumack, B. H. (Editor) Poisindex[R], Micromedex Inc., February, 1979.

Arsenic, Arsenicals:

Available Forms – Cacodylic acid and MSMA. Inorganic arsenic compounds not used in forests at present. Arsine gas forms when arsenic or arsenicals come in contact with acid, especially in presence of zinc.

Pharmacology – Arsenic inactivates sulfhydryl enzymes. Competition for sulfhydryl bindings by 2,3-dimercaprol (BAL) or d-penicillamine has resulted in reversal of arsenic poisoning. Arsine has unique effects. Depletes erythrocyte gluthathione, leading to massive intravascular hemolysis.

Clinical Effects – Early symptoms of exposure to arsenic include abdominal pain, vomiting, bloody or watery diarrhea, pain in extremities and muscles, weakness, flushing of skin and others. Chronic intoxication can occur.

Range of Toxicity – Threshold limit value of arsine gas is 0.05 ppm. Estimates of toxic levels of arsenic compounds range from 1 milligram to 10 grams, with the higher levels typical of organic arsenical herbicides. Sodium arsenite and solublized arsenic trioxide, which are not used as forestry herbicides, have estimated human mortality of 65%. Trivalent arsenic is absorbed better and is more toxic than the pentavalent form by several orders of magnitude.

Laboratory –

Blood: Blood levels normally in the vicinity of or below 0.1 ppm.

Urine: Normal output less than 50 micrograms per 24 hours, total output.

Urinary arsenic may be elevated after eating oysters.

Treatment –

1. Establish respiration; create artificial airway if necessary.
2. Prevent absorption by emesis, or, if convulsing or no gag reflex, administer endotracheal intubation preceding gastric lavage with large bore tube. Thirty to fifty grams activated charcoal in water slurry may be administered by lavage repeatedly. Sodium or magnesium sulfate, 250 milligrams per kilogram, orally may be administered as cathartic. Adults, 20-30 grams per dose up to a hundred grams total. For arsine gas, follow hematocrit and plasma hemoglobin. If plasma hemoglobin rises, institute immediate exchange transfusion. *Note:* Management of arsenic poisoning includes numerous complex procedures.

Chelation with d-penicillamine or BAL entails previous calibration of urine, and careful control. Chelation requires cardiac monitoring of BON, creatinine and liver function.

White bands may develop across the nails of hands and feet 4 to 6 weeks after exposures. Aplastic anemia may occur which is reported to respond to d-penicillamine and oxymetholone. *Note:* Organic arsenicals are of relatively low toxicity to mammals. Some exceptions have been noted in rabbits and domestic livestock. It is important to distinguish between sources of arsenic in treating intoxication. The occurrence of garlicky odor on skin and breath after exposure to organic arsenicals may indicate that metabolism is occurring. There is therefore potential for existence of some trivalent arsenicals resulting from exposure to pentavalent compounds.

Carbamate Insecticides

Available Forms – Carbaryl (Sevin). Other carbamates of higher toxicity, e.g. Temik, are not available for general forestry use. Intoxication management follows the same principles.

Pharmacology – Insecticidal and toxic effect due to reversible cholinesterase inhibition, and is enhanced by pyrethrin synergists (piperonyl butoxide, sesamex, sulfoxide, MGK264 and similar materials).

Clinical Effects – General signs and symptoms of poisoning are secondary to cholinesterase inhibition: headache, blurred vision, weakness, seating myosis, pain and/or tightness in the chest, salivation, lacrimation, nausea, vomiting, urination, abdominal cramps, diarrhea, muscular cramps and fasciculation, pulmonary adema, paraflexia, convulsions, etc.

Range of Toxicity – Carbaryl is among the least toxic of the carbamate insecticides, but a measured dose of 250 milligrams (approximately 2.8 milligrams per kilogram) of carbaryl produced moderately severe poisoning in an adult. Symptoms of poisoning develop rapidly (within minutes) following ingestion, inhalation or dermal exposure.

Laboratory – Blood cholinesterase activity is inhibited if determined by a method that minimizes reactivation (titrometric, radiometric, or colorimetric) but usual methods (Michel electrometric or delta pH method) are unsatisfactory. Cholinesterase activity determinations should be done as soon as possible after exposure if needed for diagnostic purposes.

Treatment – Establish respiration and creation of artificial airway if necessary. Morphine, thophyline, physotigmine, succinyl choline and parasympathomimetic agents are CONTRA-INDICATED IN CARBAMATE POISONING. Protopam and pralidoxine are also contra-indicated in carbaryl poisoning unless RBC-cholinesterase is reduced to 25% of normal.

Combination overdoses with organophosphates should receive treatment directed at the organophosphate itself, and #2-PAM used only if RBC-cholinesterase is less than 25% of normal.

Atropine should be administered if life-threatening cholinergic symptoms exist. A test dose of 2 milligrams intravenously is standard in an adult; in a child give .05 mg/kg. If the patient is poisoned, no signs of atropinisms will develop, and atropine may be repeated at 5 to 10 minute intervals if needed. The end point of atropine administration is cessation of secretions, and not just dilation of the pupils.

Prevent absorption by removing contaminated clothing, and washing (including hair). A single soap washing is adequate. Initial soap washing should be followed by alcohol wash and another soap wash. Tincture of green soap combines these elements.

If carbaryl was ingested, emesis should be initiated unless the patient is comatose, convulsing, or has lost the gag reflex, in which case endotracheal intubation should precede gastric lavage. Activated charcoal 5-10 times the estimated ingested dose, or 30-50 grams in a water slurry, can be administered orally or by lavage; and repeat.

Chlorinated Hydrocarbon Insecticides

Available Forms – Dicofol (Kelthane), endosulfan (thiodan), endrin, lindane, mirex, pentac, toxaphene. Dalapon – see footnote. All are efficiently absorbed by the stomach and across the skin.

Pharmacology – Principal toxic action of these higher molecular weight and complex chlorinated hydrocarbons is disruption of central nervous system function, usually manifested as respiratory depression and convulsions. Increased myocardial irritability and liver toxicity occur at extreme dosage levels, except in the case of endrin, which is strongly hepatotoxic. All of these agents induce the drug metabolizing enzymes of the liver. Some are slowly excreted, and therefore are measurable in body fat and blood long after exposure; others may be excreted in a few days or at most weeks.

Clinical Effects – Marked apprehension, hyperkinesis, vomiting, and pallor are the early manifestations of acute poisoning, especially following ingestion. When slower dermal absorption has occurred, vasciculations and myochronic movements of the extremities may be the earliest symptoms. Higher dosages bring on epileptiform convulsions accompanied by respiratory depression and interference with gas exchange from the convulsive activity. Acute symptomology nearly always appears within two hours of ingestion but interval may be longer when absorption has been by the dermal route or if dry formulations were swallowed. Hydrocarbon pneumonitis is a common complication of pesticide ingestion, because of aspiration of hydrocarbon solvent used in liquid formulations. Prognosis for full recovery is fairly good even in cases suffering convulsions. Prognosis for endrin poisonings is less favorable.

Range of Toxicity – These compounds are often less acutely toxic than the organo-phosphate compounds that have largely replaced them. Certain agents, notably endrin, are highly toxic. Toxicity is much greater when an active ingredient is dissolved in a hydrocarbon solvent. Minimal toxic dose for endrin is about twelve milligrams for the adult. Most of the other agents do not manifest toxicity at dosages below 5 milligrams per kilogram body weight.

Laboratory – Laboratories can measure concentrations of these agents in blood. Laboratories can be reached through poison control centers. Past clinical experience with many of the organo-chlorine chemicals has provided a basis for recognition of toxic concentrations in blood.

Treatment – Establish respiration and create airway if necessary. If dermal exposure, bathe and shampoo with soap and water and with rubbing alcohol or green soap to remove toxicant. Dispose of contaminated clothing in plastic bags, avoiding contact with toxicant. Convulsions may be treated with intravenous diazepam 0.1 to 0.3 milligrams per kilogram in a child, and up to a 100 milligram total dose in an adult.

Prevent absorption. If pesticide has been ingested in the past fifteen to thirty minutes, and victim is fully alert, induce emesis with syrup of ipecac. If patient is comatose, convulsing, or has lost the gag reflex, endotracheal intubation should precede gastric lavage. Activated charcoal five to ten times the estimated ingested dose, or thirty to fifty grams in a water slurry orally or by lavage will absorb pesticide. Sodium or magnesium sulfate 250 mg/kg orally given as a cathartic. Adults: 30 grams per dose to 100 grams total.

Do not administer adrenergic amines or atropine, which further increases myocardial irritability. Do not give oils which enhance absorption of toxicant. Transfusion and dialysis have not proven to be effective. Monitor myocardial, hepatic, renal, and bone marrow function in severe poisonings.

Note: Dalapon herbicide falls into this group of chemicals by virtue of being a chlorinated organic acid. Its water solubility, and extremely low toxicity, suggest that it be treated differently. As per phenoxy herbicides.

Dinitrophenol Group

Available Form – Dinoseb. Pentachlorophenol is also included in this group, and is a widely used wood preservative. Because pentachlorophenol is a widely used pesticide, and is used in wood products, it will be discussed here.

Pharmacology – This group acts as an uncoupler of oxidative phosphorylation. Exposure to toxic quantities of these materials markedly increases cellular oxidative processes. Pentachlorophenol may cause methemoglobinemia. Pentachlorophenol also contains several dioxins, which may cause skin disorders such as chloracne. Porphyria subcutanea tarda has been associated with dioxin poisoning. Dioxin concentrations in pentachlorophenol may be toxicologically significant.

Clinical Effects – Hyperactivity with convulsions leading to coma are preceded by nausea, fever and extreme restlessness. Sinus tachycardia is seen initially and in severe overdoses. Ecotopic foci will be noted as well as ventricular arrhythmias, including ventricular tachycardia progressing to ventricular fibrillation. Hyperpyrexia is usually seen with rapid respiratory rate, diaphoresis and pulmonary edema. Concentrated material ingested may produce burns of the oral and pharyngeal mucosa. Esophageal burns have not been noted. Renal failure may occur twelve to seventy-two hours after exposure. Jaundice may be noted as early as two days following ingestion. Lethargy and weakness are noted when recovery begins.

Range of Toxicity – Onset of symptoms is so rapid that if a patient has not demonstrated any within a few hours following ingestion, it is unlikely that he will ever do so. The toxic dose in an acute state is probably in excess of 3-5 mg/kg. Inhalation is a source of marked toxicity and is usually encountered when the material is used as a spray, or as a wood preservative in enclosed areas.

Laboratory – Parent drug can usually be detected in plasma and urine, but is of little value in the acute management of such patient. Ten micrograms per milliliter (10 ppm) is probably the lowest limit of toxicity.

Treatment – Establish respiration and clear airway. Administer oxygen for cyanosis. Prevent absorption, by emesis unless the patient is comatose, convulsing, or has lost the gag reflex, in which case endotracheal intubation should precede gastric lavage. Activated charcoal 5 to 10 times the estimated ingested dose, or 30 to 50 grams, in a water slurry, orally or by lavage, may be repeated to absorb the chemical. Sodium or magnesium sulfate, 250 mg/kg orally, may be given as a cathartic. Adults may receive 20-30 grams per dose, to 100 grams total.

Exposed areas should be decontaminated with soapy water by medical personnel wearing protective plastic or rubber clothing. Cutaneous exposure to dinitrophenols (not pentachlorophenol), is usually accompanied by yellowish discoloration which does not have to be removed in its entirety to prevent absorption. Induction of hypothermia with tepid water baths is superior to alcohol sponging or ice packs which tend to constrict peripherally.

Administration of salicylates to reduce hyperpyrexia is contra-indicated. These drugs are also oxidative phosphorylation uncouplers. Fluids, electrolytes, and acid-base data should be monitored and losses replaced. Intravenous solutions should contain adequate amounts of glucose to supply the requirements of increased metabolism.

Hydrocarbons (Petroleum Distillates, Diesel, etc.)

Available Forms – Available in numerous products and listed as non-toxic components of others. Many poisonings are due, not to a product component, but to the aspiration of the "non-toxic" hydrocarbon solvent.

Pharmacology – Toxicity of hydrocarbons is usually related to viscosity. Motor oils and heavy greases have very limited toxicity. Low-viscosity materials such as mineral seal oil have high aspiration hazard. Small amounts of low-viscosity material (gasoline, kerosene, solvent), aspirated can spread over a large portion of the pulmonary bed resulting in chemical pneumonitis.

Clinical Effects – The patient may begin by choking, coughing, and gagging as he swallows; shortly the patient may begin vomiting following ingestion, and develop diarrhea as well. Aspiration of the products at the time of initial swallowing may result in aspiration pneumonitis on a chemical basis. Central nervous system depression may occur. Cyanosis, tachycardia, and hyperpyrexia may appear as a result of chemical pneumonitis. Albuminuria, hematuria, heptatic damage and cardiac arrhythmias including ventricular fibrillation may be seen. Inhalation of petroleum distillates may cause a euphoria-a transient high-associated with later headaches, nausea and sometimes dispnea. The patient may become confused and apathetic. Chronic inhalation may result in hepatic and renal damage. Death may be due to vent fibrillation, respiratory arrest or anoxia, secondary to replacement of atmosphere surrounding the patient is lung tissue with hydrocarbon.

Rash may occur due to de-greasing effect of many hydrocarbons on skin surface. Symptoms of cerebellar dysfunction, particularly chorea and ataxia, have been reported from sniffing leaded gasoline. Lead poisoning may also be a result of ingestion of gasoline.

Range of Toxicity – It is difficult to estimate amounts of hydrocarbons capable of producing severe symptomology. Less than 1 cc of some, when aspirated directly into the trachea, may produce severe pneumonitis and death. It is impossible to evaluate the amount ingested by a child found with an empty bottle surrounded by a pool of hydrocarbon.

Laboratory – Identification of various hydrocarbons in blood and other tissues is available in some laboratories.

Treatment – Establish respiration, create artificial airway if necessary. Epinephrine is contra-indicated. In the presence of hydrocarbons ventricular fibrillation may result. Convulsions may be treated with intravenous diazepam 0.1 to 0.3 mg/kg in a child and up to 10 milligrams in an adult.

If the patient is already coughing on arrival at the medical facility, aspiration has already occurred. Do not induce emesis in a patient who is comatose, convulsing, or has lost the gag reflex.

Emesis is indicated for some hydrocarbons and not for others. The following examples are treated with emesis because of their inherent central nervous system toxicity or their potential for toxic effects: halogenated hydrocarbons, aromatic hydrocarbons such as toluene, xylene, benzene, or turpentine if doses of 1 milliliter per kilogram or more are ingested. Gasoline, kerosene, lighter fluid, petroleum ether, petroleum naptha, stoddard solvent, mineral turpentine, or petroleum spirits if doses are one milliliter per kilogram or more ingested, with a minimum of one ounce or 30 milliliters dosage. Any hydrocarbon or petroleum distillate in any amount containing dangerous additives, heavy metals, insecticides, etc.

Agents where vomiting is not recommended include mineral

seal or signal oil, such as found in furniture polish or oil polishes. Agents considered generally non-toxic include asphalt, tar, lubricants, mineral oil, or fuel oil such as diesel.

Cathartics should follow all ingestion of hydrocarbons. Mineral or vegetable oil carthartics should be avoided. The concept of cautious gastric lavage is outdated. Lavage should be reserved for patients with absent gag reflex, coma, or convulsions. It *must* be preceded by intubation, with a cuffed endotracheal tube if emesis cannot be done. Analeptics are contra-indicated in central nervous system depression resulting from hydrocarbons, just as they are in depression from anything else.

Lacrimators (Soil Fumigants, etc.)

Available Forms – Chloropicrin (trichloronitromethane) is a lacrimator which has been used by the military as a tear gas agent and also as a fumigant insecticide and as an ingredient in mixed insecticides to warn of impending toxicities of the other chemicals. It is often mixed with methyl bromide, which is more toxic.

Pharmacology – Chloropicrin has the toxic effect of interfering with oxygen transport by its reaction with SH-groups in the hemoglobin. It is possible there may also be a photochemical transformation of chloropicrin into phosgene.

Clinical Effects – Lacrimation is the most important clinical effect and depending upon the concentration in the atmosphere, may last 15 to 30 minutes or less. Second exposures may cause less lacrimation. Some redness and edema may be noted for a day or two following exposure to any of these agents. Laryngospasm may occur in response to the irritant effects. In addition to the lacrimation effects of chloropicrin, vomiting is very common in those exposed to this particular chemical. All body surfaces exposed will be irritated by chloropicrin.

Range of Toxicity – The threshold limit value (TLV) for chloropicrin is 0.1 ppm. The lethal dose is 5 ppm for 10 minutes; when breathed for 1 to 2 minutes at 20 ppm, there will be definite bronchial or pulmonary lesions.

Laboratory – The materials are generally undetectable because the concentrations are low.

Treatment –

1. Establishment of respirations and creation of an artificial airway if necessary. Check adequacy of Tidal Volume (Normal 10-15 cc/kg).

2. Remove the patient from the contaminated atmosphere as rapidly as possible.

3. Flood the eyes with water or saline for at least 5-15 minutes. This procedure should be performed with a large volume of water, gently poured into the eye. It should NOT be done with a small eye dropper at a low rate. The concept here is to flush the eyes and dilute out the toxic material as quickly as possible.

4. Laryngospasm may require intubation to permit adequate breathing.

5. Removal and decontamination of clothing should be carried out to avoid further discomfort. This may generally be performed by washing the clothes and allowing non-washable items to air for a few days.

6. Ophthamologic evaluation should be performed after decontamination has been completed.

7. Patients arriving at a hospital in an ambulance or other conveyance will most likely have contaminated the vehicle and it will need to be decontaminated with soap and water wash to avoid discomfort to other passengers.

8. Patient's clothing should be placed in large plastic bags and sealed to avoid discomforting vapors and contamination of emergency room personnel.

9. Emergency room personnel, when notified of the possibility, should wear aprons, rubber gloves and masks to avoid becoming contaminated.

Organophosphates

Available forms – Azinphosmethyl (guthion), diazinon, dimethoate, disulfoton (Di-syston), fenitrothion (Sumithion), gardona, imidan, malathion, Metasystox-R, orthene, phosphamidon (dimecron), thimet, trichlorfon (Dylox). Many of these compounds are formulated as solutions in petroleum distillates.

Pharmacology – The widely used organophosphate insecticides inhibit (phosphorylate) blood-borne and tissue cholinesterase enzymes, including the acetyl cholinesterase located at para-sympathetic neuroeffector junctions, ensympathetic ganglia, and in brain tissue. Organophosphates are efficiently absorbed from the gut and across the skin. Most are hydrolyzed in the body to alkyl phosphates and various "leaving groups" which determine toxicity. Accumulated acetylcholine in critical tissues causes persistent stimulation of end organs (cholinergic crisis). Atropine blocks the effects of excessive acetylcholine at receptor sites so long as the atropine persists in high concentration. Certain oxime compounds regenerate functional cholinesterase if phosphorylation of the enzyme has not persisted more than 36 hours.

Clinical effects – Prodomal symptoms: headache, dizziness, nausea, Muscacinic manifestations: salivation, vomiting abdominal pain, diarrhea, chest pains, bradycardia, bronchorrhea, dispnea, miosis, blurred vision, diaphoresis. Nicotonic manifestations: muscle twitching and spasms, profound weakness. Central nervous system manifestations: psychotic behaviour, incoordination, unconsciousness, rarely convulsions. Protracted low-level absorption may produce a syndrome similar to influenza. High dosage may cause a toxic psychosis similar to acute alcoholism. Has been misdiagnosed as asthma and congestive heart failure. Rarely may cause peripheral neuropathy.

Range of Toxicity – Organophosphate insecticides have a wide range of toxicity, but include the most toxic pesticides used in forestry (LD_{50}, 2-1, 375 mg/kg). Acute toxicity depends strongly on kinetics of absorption, e.g. sudden absorption of low toxicity compounds may have a grave effect.

Laboratory – Nearly all organophosphates depress the activities of either plasma pseudocholinesterase or red cell acetyl cholinesterase, or both. Usually this biochemical manifestation of toxicity appears at lower dosage levels than amounts producing symptoms or signs. Comparison with pre-exposure value is helpful. Plasma cholinesterase usually recovers in a few days or weeks: RBC-acetyl cholinesterase recovers in several days to eight months depending on severity of depression. Urinalysis for alkyl phosphates is a sensitive estimate of exposure.

Treatment – Establish respiration and maintain clear airway. In cases of dermal exposure, bathe and shampoo with soap and water, and with rubbing alcohol or green soap to remove toxicant. Dispose of contaminated clothing in plastic bags, avoiding contact with toxicant.

Prevent absorption by emesis unless the patient is comatose, convulsing, or has lost the gag reflex, in which case endotracheal intubation should precede gastric lavage. Activated charcoal 5-10 times the estimated ingested dose, or 30-50 grams in a water slurry, orally or by lavage, will absorb chemical. Use sodium or magnesium sulfate 250 mg/kg orally as a cathartic. Adults: 30 grams per dose to 100 grams total.

Draw blood for measurement of cholinesterase. Through the

same needle inject atropine sulfate slowly: 2-5 milligrams adult dose, 0.05 mg/kg for children under twelve. Repeat dosage every 10-30 minutes or as needed to attain and maintain full atropinization, best indicated by complete clearing of bronchial and pulmonary rales. Atropinization must be maintained until absorbed toxicant has been metabolized. This may require 2-2,000 milligrams (or more) atropine over a few hours or several days. In severe poisoning, atropine must be withdrawn cautiously to avoid remission, often in the form of pulmonary edema.

Severe poisoning, particularly those characterized by profound weakness and muscle twitching, should be treated with pralidoxime in addition to atropine. Administer 1.0 gram intravenously at 0.5 grams per minute, or infuse in 250 ml saline over 30 minutes.

Do not give adrenergic amines, amino phylline, succinylcholine, phenothiazines, or reserpine or oils which enhance absorption. Be prepared to treat hydrocarbon pneumonitis.

Convulsions occur rarely in organophosphate poisoning. Convulsions may be treated with intravenous diazepam 0.1 to 0.3 milligrams per kilogram in a child, and up to 100 milligrams in an adult. Monitor for respiratory arrest or hypotension.

Veterinary dosages: Atropine: dog and cat 0.1 to 0.2 mg/kg; cattle 0.5 to 1.0 mg/kg; horse 0.1 to 0.2 mg/kg. Repeat as frequently as needed to get atropine effect, except in cattle in which effects last one to two hours. Pralidoxine: dog and cat 20 mg/kg; cattle 20 mg/kg; horse 4 mg/kg. May repeat in 1 hour if necessary.

Phenoxy herbicides

Available forms – Free acids of these compounds, (2,4-D, dichlorprop, 2,4,5-T, Silvex) as well as various esters and salts are formulated in water suspensions (sometimes solutions) or in various organic solvents. Some of the salt formulations may be strongly basic. Some esters are fairly volatile, salts are not.

Pharmacology – Chlorophenoxy herbicides are moderately irritating to skin and mucous membranes of workers heavily exposed. They are absorbed across the gut and weakly across the skin, and are excreted intact within hours or at most a few days mainly by the kidneys. Given in large dosages to dogs, they cause ulcers of the mouth and pharynx, and myotonia of the hind extremities – a stiffness of movement caused by reversible injury to the spinal cord. They are moderately toxic to liver and kidneys. LD_{50} values in laboratory animals are in the 300-1,000 mg/kg range. In common with many other chemicals, 2,4,5-T and the contaminant TCDD (and perhaps 2,4-D) are teratogenic in some laboratory animal species. Degeneration changes in brain cells are reported in a single suicidal ingestion case (2,4-D). Present formulations of 2,4,5-T contain so little TCDD that TCDD effects, if any, are likely to be masked by those of 2,4,5-T. Contamination of food by TCDD has not been reported at toxicologically significant levels.

Clinical effects – Irritation of skin, eyes, nose and throat occurs in handlers, usually resolving in a few days after contact ceases. Vertigo, headache, malaise and parathesias have been reported occasionally in applicators. Sensitization has not been reported. On ingestion, burning of the mouth, esophagus and stomach is prominent, often leading to nausea, vomiting and diarrhea. Muscle twitching and spasms occur at high dosage rates, followed by profound muscle weakness and stupor. Convulsions have occurred antemortem in one suicide.

Albuminuria, elevated LDH and SGOT reflect moderate degrees of kidney and liver injury. In a single reported case of 2,4-D ingestion, the victim developed hyperthermia, rabdomiolysis, and acute tubular necrosis, secondary to myoglobinemia. Chloracne from chlorodioxins has occurred in manufacturing plant workers, but not in handlers of the commercially packaged herbicide. Occasional cases of protracted neuropathy have been reported following what seemed to be slight dermal exposure. On the other hand substantial doses have been ingested by volunteers and injected intravenously into patients without neuropathic effects or other serious symptomology, unless extreme doses were given. Individual idiosyncrasy may therefore be involved in the rarely reported neuropathies.

Laboratory – These herbicides can be measured in plasma and urine by gas-liquid chromatography. Chlorophenoxy compounds do not affect blood cholinesterase activities.

Treatment – Establish respiration and maintain airway. Decontaminate skin by washing with green soap and water. Flush contaminated eyes with copious amounts of water. Prevent absorption by emesis unless the patient is comatose, convulsing or has lost the gag reflex, in which case endotracheal intubation should precede gastric lavage. Activated charcoal 5-10 times the estimated ingested dose, or 30-50 grams in a water slurry are given orally or by lavage. Sodium or magnesium sulfate, 250 mg/kg orally, is given as a cathartic. Adults: 30 grams per dose to 100 grams total. Convulsions may be treated with intravenous diazepam 0.1 to 0.3 mg/kg in a child and up to 100 mg in an adult. Be prepared to combat respiratory depression and hypotension. Test urine for protein, RBD, and myoglobin. Induce alkaline diuresis if myoglobin is present. Chemodialysis is not likely to be effective. Maintain adequate urine flow. Manage hypothermia with sponge baths.

Pyrethrum, Allethrin, Pyrethroids:

Available forms – Pyrethrum is a very old insecticide which is obtained by grinding or extracting the flowers of the pyrethrum plant, *Chrysanthemum cinerafolium.* The flower heads contain 1 to 3% active ingredients (pyrethrolone and cinerolone) whose esters are called pyrethrin I and II and cinerin I and II. Contained in numerous household sprays. Also called Dalmation Insect Flowers, Trieste Flowers and Persian Insect Powder. Allethrin is the allyl homolog of cinerin I. Also called Pyresin, Pynamin and Pallethrine. Bloresmethrin is a related pyrethroid which is being used as the starting material for a variety of new synthetic pyrethroid derivatives with enhanced insecticidal activity. The pyrethroids are commonly formulated in deodorized kerosene with piperonyl butoxide or Sesamin (Insecticidal synergists) for liquid applications, in a volatile liquid (dichlorofluoromethane or methyl chloride) for aerosol applications or as an impregnated dust. They have low persistence.

Pharmacology – The pyrethroids are powerful contact insecticides which cause rapid paralysis (quick knockdown) in house flies and many insects and appear to act on the neuromuscular system. Pyrethrum and allethrin have low toxicity to warm-blooded animals but some of the newer synthetic pyrethroids, such as resmethrin and tetramethrin, are considerably more toxic. Pyrethrum dermatitis in sensitized individuals is thought to be due to the presence of a volatile oil present in the flower heads of the plants. Toxic doses in animals produce diarrhea, clonic convulsions, prostration and respiratory paralysis. Rapidly detoxified by hydrolysis which is inhibited by synergists such as piperonyl butoxide.

Clinical effects – Dermatitis in sensitized individuals is most common reaction. More common in individuals sensitive to ragweed pollen. Death due to anaphylactic shock has been reported. Severe exposure may cause hyperexcitability, incoordination, tremors, muscular paralysis, numbness of the lips and tongue, sneezing, nausea and vomiting. Bronchospasm may occur from inhalation. Chronic inhalation resulting in hypersensitivity pneumonitis with malaise, shortness of breath, pleuritic chest pain and cough has been reported.

Range of Toxicity – Lethal dose for man estimated on basis of animal studies is more than 1 to 20 grams orally. More toxic parenterally. Petroleum solvents used as vehicles are generally more hazardous than the pyrethrums or the synergists found in common household sprays. Severe poisoning rare.

	Acute Oral (mg-kg)
Pyrethrin	200-2600
Allethrin	680-1000
d-trans Allethrin	425-575
Resmethrin	1500
Tetramethrin	20,000

Laboratory – Color test with 2-2 (2-aminoethylamine) ethanol produces red to violet color (Feinstein, L., Science 115:245, 1952)

Treatment –

1. Establishment of respirations and creation of an artificial airway if necessary. Check adequacy of Tidal Volume (Normal 10-15 cc/kg).

2. Prevention of absorption:
   a. Emesis should be initiated unless the patient is comatose, convulsing or has lost the gag reflex.
   b. If these contraindications are present, endotracheal intubation should precede gastric lavage with large bore tube.
   c. Activated charcoal 5-10 times the estimated ingested dose or 30-50 grams in a water slurry orally or by lavage.
   d. Sodium or magnesium sulfate 250 mg/kg orally as a cathartic. Adults: 30 grams/dose to 100 grams total.

3. Convulsions may be treated with intravenous diazepam 0.1-0.3 mg/kg in a child and up to 100 mg in an adult.

Rotenone:

Available forms – Derrin, Tubatoxin, Nocouline, Derris, Cube, Barbasco, Halari, Nekos, Timbo. Noxfish (5% rotenone), Pro-Noxfish (2.5% rotenone plus 2.5% sulfoxide).

Pharmacology – A selective contact insecticide with some acaricidal action. Low persistence in spray or dust residues. Rotenone is a crystalline neutral principle which is the active constituent in the roots of *Derris elliptica* and lonchocarpus plants. Used mainly in dusts and sprays alone and in combination with other insecticides as washes and dips against animal ectoparasites, as fish toxicants, in household sprays and in the treatment of chigger bites and scabies. It is rapidly oxidized to less toxic compounds by light and alkali so dust formulations are often stabilized by the incorporation of small amounts of acid (phosphoric acid, etc.). Synthetic derivatives of rotenone (isorotenone, dihydrorotenone) are more insecticidal but not in common usage. Rotenone inhibits mitochondrial oxidation involving NAD in a manner similar to amobarbital and this effect can be antagonized with Vitamin K2 but not with 2,4-dinitrophenol, NAD or cytochrome C. Rotenone has spasmolytic effects like those of papaverine. More toxic parenterally than orally (due to poor absorption). Oily cathartics enhance absorption. Fresh derris root more toxic than dry.

Clinical effects – Local effects include conjunctivitis, dermatitis, pharyngitis and rhinitis. Inhalation of dust can produce severe pullmonary irritation. Ingestion results in gastointestinal irritation, nausea and vomiting, gastric pain, and numbness of the oral mucosae. It is reported to have a stimulant effect on the emetic center and a transient stimulant effect on respiration.

Range of toxicity – Relatively non-toxic for most warm-blooded animals because of poor absorption and prompt emesis. Highly toxic to swine and fish. Lethal dose in man is estimated from animal data to be about ½ gram.

Laboratory – Hydriodic acid produces a characteristic light blue color with rotenone (Delfel: J Agric. Food Chem. 13:56, 1965).

Treatment –

1. Establishment of respirations and creation of an artificial airway if necessary. Check adequacy of Tidal Volume (Normal 10-15 cc/kg).
2. Prevention of Absorption:
   a. Emesis should be initiated unless the patient is comatose, convulsing or has lost the gag reflex.
   b. If these contraindications are present, endotracheal intubation should precede gastric lavage with large bore tube.
   c. Activated charcoal 5-10 times the estimated ingested dose or 30-50 grams in a water slurry orally or by lavage and repeat.

d. Sodium or magnesium sulfate 250 mg/kg orally as a cathartic, Adult: 20-30 grams/soe to 100 grams total

3. Convulsions may be treated with intravenous diazepam 0.1-0.3 mg/kg/dose in child and up to 10 mg in an adult.

4. Administer glucose intravenously to correct hypoglycemia if present.

5. Use of menadione sodium bisulfite 10 mg/IV may be indicated to counteract cellular respiration depression and hypotension in severe cases.

Sulfur compounds (Lime sulfur)

Available forms – Sulfur is used in manufacturing sulfuric acid, as an insecticide and fungicide, and various other processes. Sulfur Lime (Calcium polysulfide), sodium sulfide, ammonium sulfide and thioacetamide release hydrogen sulfide in contact with water or acids. Sulfur dioxide (SO_2) is found in a compressed or cold solution since it has a boiling point of minus ten degrees.

Pharmacology – Sulfur has few systemic actions. In the body it is changed partly to hydrogen sulfide. Ingestion may cause hydrogen sulfide production due to bacteria in the colon. Sulfur dioxide reacts to produce sulfurous acid when it reacts with water or mucous membranes or on alveolor membranes.

Clinical effects – Hydrogen sulfide odor on breath (rotten eggs), vomiting, difficulty swallowing, fever, redness of tongue and pharynx, headache, vertigo, excitement or depression, loss of memory and prostration, abdominal pain, diarrhea and urinary disturbances may occur. Tremors, convulsions and death may result. Pneumonia and peripheral neuritis may follow after recovery. Lime sulfur is irritating on contact with skin, eyes and mucous membranes. Carbon disulfide damages the CNS, the peripheral nerves and the hemopoletic system. It causes congestion and edema in the GI tract. The mercaptans in high concentrations cause cyanosis, convulsions, hemopoletic system. It causes irreversible depression of the CNS function. Metabolic acidosis may occur.

Range of Toxicity – Ingestion of 10-20 grams or absorption through the skin may cause signs of hydrogen sulfide poisoning. The lethal dose of soluble sulfur has been reported from 10 grams to 0.5-5.0 grams/kg. The TLV is 60 mg/cu.M, or 20 ppm, the approximate fatal dose by ingestion is 1 gram.

Treatment –

1. Establishment of respirations and creation of an artificial airway if necessary. Check adequacy of Tidal Volume (Normal 10-15 cc/kg).
2. Prevention of Absorption:
   a. Emesis should be initiated unless the patient is comatose, convulsing or has lost the gag reflex.
   b. If these contraindications are present, endotracheal intubation should precede gastric lavage with large bore tube.
   c. Activated charcoal 5-10 times the estimated ingested dose or 30-50 grams in a water slurry orally or by lavage and repeat.
   d. Sodium or magnesium sulfate 250 mg/kg orally as a cathartic, Adult: 20-30 grams/dose to 100 grams total.
3. For skin and eye contact (lime sulfur), remove with copious quantitites of water.
4. Treat pulmonary edema if present – usually will require positive and expiratory pressure and steroids may be useful.

Triazines

Available forms – Both symmetrical and asymmetric triazine compounds are used in forestry. Atrazine and simazine are symmetric triazines; hexazinone is an asymmetric triazine used in similar fashion. Many other triazines are used in agriculture. Amino triazole has similar toxicological behaviour, although it is unrelated to the triazines herbicidally. These compounds range in water solubility, with simazine being nearly insoluble in water, and amitrole being completely soluble in water. None are soluble in oil or oil solvents. Some agricultural triazines are suspended in organic solvents, permitting them to emulsify in water.

Pharmacology – The triazines are well absorbed by the mammalian gut, and probably across the skin. Laboratory animal studies and occupational exposure experience both indicate low systemic toxicity. Following high chronic dosage in rats, some triazines cause growth retardation, slight leucopenia, and modification of liver metabolism. Extreme acute doses cause edema of lung and brain, generalized hemorrhage, broncho-pneumonia, nephrosis and liver injury. Sheep and cattle exhibit neuro-muscular disturbances. Although workers in some triazine manufacturing plants suffer skin irritation from exposure to the primary reactants, the finished triazines have not caused significant skin or mucus membrane irritation in regular handlers.

Clinical effects – Effects in man have not been reported. When fed to sheep and cattle at high dosage, atrazine causes anorexia, depression of activity, muscle spasms and fasciculations, ataxia and dyspnea.

Laboratory – The triazines can be measured in blood and urine by properly equipped laboratories, which can be reached through poison control centers.

Treatment – If persons exposed to triazines exhibit symptoms and signs of severe toxicosis, concurrent absorption of other or additional toxicants should be suspected. Careful inquiry should be made into the full spectrum or recent exposure so as not to overlook more serious toxic hazards.

Wash contaminated skin with soap and water. Flush contaminated eyes with copious amounts of water. Establish and maintain respiration. Prevent absorption by emesis unless the patient is comatose, convulsing or has lost the gag reflex, in which case apply endotracheal intubation preceding gastric lavage. Activated charcoal at 5-10 times the estimated ingested dose or 30-50 grams total may be used in a water slurry orally or by lavage. Sodium or magnesium sulfate 250 mg/kg may be administered orally as a cathartic. Adults 30 grams per dose to 100 grams total. Monitor CBC, urine for albumin and cells, liver enzymes (alkaline phosphate, LDH and SGOT).

General or Unknown

General – This management is intended for use in the absence of a specific treatment protocol for a product, drug, or chemical. Some old products such as ammonium sulfamate, and DCPA fall into this category. There is a possibility that new chemicals may not be listed and in the current files of poison centers, and some guidelines may be needed for patient care. Experimental agents are sometimes not available for listing and may be ingested.

Initial patient contact – General – establish respiration and create airway if necessary. Treat hypotension with fluids or plasma – try to avoid vasopressors if possible. Note: There are a few substances with a luxury of an antidote – good patient support with maintenance of vital signs is critical. Respiratory assistance and treatment of hypotension are not contra-indicated in any known ingestion.

Evaluation of Toxicity – What clinical findings are demonstrated by the patient? What class of compounds are they consistent with? E.G., a despondent patient in coma most likely has taken a sedative, hypnotic, or opiate. Further examination may demonstrate tachycardia, large pupils, atony of the bowel, and fever leading to consideration of an ingestion with anticholinergic properties such as over-the-counter sleeping preparations, "cold" medications or tricylic antidepressants. *Treat the patient, not the poison!*

Determine if specific information is provided by the manufacturer on the label or in accompanying literature, consult with the analytical toxicology laboratory.

Coma levels

0 = Asleep but can be aroused and answer questions.
1 = Comatose, does not withdraw from painful stimuli, reflexes intact.
2 = Comatose, does not withdraw from painful stimuli, most reflexes intact. No respiratory or circulatory depression.
3 = Comatose, most or all reflexes are absent but without depression of respiration or circulation.
4 = Comatose, reflexes absent, respiratory depression with cyanosis, circulatory failure or shock.

Coma should be treated in an appropriate manner regardless of the suspected cause. Attention to breathing and shock should be paramount. The use of naloxzone is suggested in a suspicious coma, .005 mg/kg. It does not cause respiratory depression even if there is no opiate drug on board. It does not act in synergism with barbituates.

Specific therapy – Emesis – although vomiting with a syrup of Ipecac produces a mean recovery of no more than 30% of an ingested dose, it is still probably a useful procedure. Save initial emesis for analysis.

Contra-indications to emesis: ingestion of strong acids or bases – look for burns of the oral mucosa. Prevent absorption by emesis unless the patient is unconscious, has lost the gag reflex, or is seizing.

Petroleum hydrocarbons – While thought in the past to be a contra-indication, emesis is now encouraged by most poison centers. See petroleum distillate management.

Dosage for emesis: 1) Ipecac – the most readily available drug for induction of vomiting is a syrup of Ipecac (not the elixir or fluid extract which is toxic). Dose 15 cc (half-ounce, one tablespoon) orally repeated on time, if necessary, for children; 30 cc (one ounce, two tablespoons) orally for adults. Follow with large amounts of fluids (one-two glasses). Ipecac on an empty stomach is like squeezing an empty balloon. Keep the patient ambulatory. After fifteen minutes, stimulate the patient's throat to help emesis.

Lavage – While lavage was considered inferior to emesis for many years, this is no longer the case. The difference has been the use of the 36-50 French orogastric hose, and copious amounts of fluid. Tap water can be safely used in adults although saline should be used in young children. Save initial aspirate for analysis. Cautions: If the patient is unconscious, seizing, or has lost the gag reflex, endotracheal intubation should precede lavage. Other patients may be lavaged without intubation if placed in the head-down left side position, and have suction available.

Charcoal – Charcoal is not a universal antidote. Many agents are adsorbed. In any case the charcoal will act as a marker of intestinal motility and provide knowledge that gastrointestinal transit time has occurred and further absorption of drug is unlikely, or the gut has become cleared. Many herbicides are adsorbed by charcoal.

Cathartics – These agents move ingested material through the GI tract rapidly and prevent continued absorption. Oil-based cathartics are questionable in their use since they may cause aspiration pneumonia. Chemical cathartics are equally effective.

Contra-indications: magnesium-containing cathartics and renal failure, oil-based cathartics and pesticide poisoning. Dose: sodium or magnesium sulfate 250 mg/kg orally as a cathartic. Adults, 20-30 grams per dose to a 100 grams total.

Decontamination – Many agents, especially insecticides, can be rapidly absorbed through the skin and may continue to cause symptoms with prolonged absorption. Others such as gasoline which causes skin burns on contact, and acids or caustics which cause burns as well, need to be removed from the skin. The safest decontaminant is water with soap. Use copious amounts gently, and try not to abrade the skin. If there is any question of exposure, skin decontamination should be thorough.

Eye Exposure – Decontamination should occur at home or at work with plain tap water. Treat immediately with irrigation and continue for 20 minutes. Irrigation means holding the head back over the sink and pouring saline or plain water from a large pitcher or liter bottle into the affected eye. Irrigation should dilute or remove the agent quickly. The first person who sees the patient should perform the irrigation immediately. Consult with opthalmologist.

Forced diuresis – Forced diuresis is frequently used in serious poisonings if the drug is excreted in the urine in active form. It should not be used unless it is specifically indicated, as it may increase the problem of cerebral adema, a common complication in the poisoned patient.

Laboratory – Specific laboratory tests and screens are available for compounds and groups of compounds. It is critical to consult the analytical chemist early so that correct samples may be obtained for the suspected poison. Discussing symptomology and suspicions from history will help the analyst reach an answer more quickly. Sending specimens with a request for poisoning screen will take the chemical analyst many hours and prolong identification so that definitive care may be delayed. Supportive care should be provided while laboratory results are pending. Specimens should be taken from gastric contents by saving all emesis produced and sending entire amount to laboratory. When using lavage, save initial aspirate separately from the large amount of lavage solution, but send both. For blood, discuss sample containers and amounts of blood with the analyst, but in general 10 cc of heparinized blood will suffice.

Note: Leather absorbs pesticides and cannot be decontaminated. Leather should not be worn while handling pesticides, and contaminated leather should be destroyed.

GLOSSARY

Absorption

(Phys. liq.) The movement of a liquid into a solid, as if through pores. Soaking in.

Active Ingredient

(Pest.) The parent substance that causes the pesticidal activity in a product. Activity enhancers and formulation components are not regarded as active ingredients unless they have pesticidal activity when administered alone. When an acid is reacted with an alcohol to form an ester, it is the acid equivalent in the ester that is used as the basis for active ingredient designation.

Acutely Toxic

(Toxicol, Pest.) Having the ability to produce toxic effects shortly after administration of a certain dosage. The term applied to toxicants that produce toxic effects at one level of a single dosage, but whose effects are tolerated with repeated doses of somewhat lower level. Generally, acutely toxic substances are eliminated quickly or metabolized to inert substances by animals. Nearly all herbicides and non-chlorinated insecticides are in this group.

Acute Toxicity

(Toxicol.) Toxicity of a substance manifested by symptoms occurring within a few days of ingestion. Acutely toxic does not imply highly toxic, which is a matter of *degree* of toxicity. If a substance is acutely toxic, symptoms appear early after exposure; recovery is usually rapid.

Adjuvant

(Pest.) A substance added to a pesticidal formulation or spray mix to enhance economic usefulness, drift control and other technical properties.

Adsorption

(Phys.) The binding of molecules to a surface by chemical energy. Adsorption is an equilibrium reaction, reversible.

Agitation

(Pest.) The physical churning of a liquid mixture to prevent settling out of insoluble ingredients.

Air Blast Machine

(See Mist Blower)

Amine

(Chem.) A compound derived from ammonia, with one or more hydrogens replaced with an organic group. Commonly used to form water-soluble salts of certain herbicides. As the length of the carbon chains increase, solubility in oil increases. A 12-carbon amine is used as an oil-soluble formulation.

Amino Acid

(Chem.) An organic acid containing an amino group on the beta carbon. Amino acids are the basic building blocks for all proteins.

Arsenical

(Chem.) An arsenic compound.

Asymmetric

(Chem.) Non-symmetrical about any axis. For example, atrazine is a symmetrical (s-) triazine because the triazine ring has uniformly unsaturated structure and symmetrical distribution of carbon and nitrogen atoms. Hexazinone has a similar arrangement of N and C atoms, but only one unsaturated bond and non-planar configuration. It cannot be divided into two equal parts along any plane, hence is asymmetric.

Bark Beetle

(Silv.) An insect that feeds and lays eggs in the phloem of woody plants. Order Coleoptera.

Basal Spray

(Pest.) Application of a pesticide, usually a herbicide for controlling brush or weed trees, directed at the base of the stem. Most basal sprays are phenoxy or other ester preparations in an oil carrier.

BE Ester

(Pest.) Butoxy ethanol ester. One of the low volatile esters, usually of phenoxy herbicides.

Biocide

(Pest.) A toxicant producing effects in all forms of living organisms in a more or less comparable range of exposure.

Boom

(Pest.) The part of a sprayer on which multiple nozzles are mounted.

Broadleaf Weed

(Bot.) Any dicotyledonous plant growing where it is not wanted.

Bipyridinium Ion

(Chem., Pest.) A compound consisting of two 6-member rings, each having a nitrogen in the 4 position, connected by a carbon linkage. Includes paraquat and diquat herbicides.

Canopy

(Silv.) The upper layer of a forest containing most of the leaves. Sometimes refers to understory layer as well.

Carbamate

(Chem., Pest.) Any compound having an R-N-COOH structure. Includes a large group of insecticides and herbicides.

Carcinogen

(Toxicol, Pest.) A substance that causes formation of malignant tumors in animals.

Carrier

(Pest.) A substance used to dilute a pesticidal product for ease of distribution or enhancement of effect.

Certified Applicator

(Pest.) One who has undergone training and taken a formal

examination, the passage of which results in certification. There are numerous specialties and several degrees of difficulty according to the type of certification.

Chemical Thinning

(Silv.) The use of herbicides to achieve desirable spacing and structure among desirable species, usually by injection.

Chlorinated Hydrocarbon

See organochlorine –

Cholinesterase

(Biol.) An enzyme involved in the transfer of nerve impulses to muscles.

Chronically Toxic

(Toxicol, Pest.) Having the ability to produce toxic symptoms with repeated dosages, any one of which will not produce symptoms. Toxicants that accumulate in tissues or those that inhibit repair mechanisms, such as organochlorine insecticides and TCDD are in this group.

Chronic Toxicity

(Toxicol.) Toxicity of a substance manifested by symptoms occurring as the result of a long period of low-level exposure and accumulation of elevated body burden of the substance. Chronically toxic substances produce symptoms which tend to persist for longer than those of acute intoxicants.

Cleaning

(Silv.) The removal of competing vegetation around desirable young trees.

Common Name

(Pest.) Name assigned to the active ingredient of a pesticidal material by one of the responsible scientific organizations. Usually shorter than the chemical name, the common name is used to facilitate communication. A single common name is assigned to each compound, and is used universally in the scientific literature to describe it.

Conjugation

(Chem.) The combination of two compounds by molecular attraction. Properties of both compounds are affected. Sometimes reversible. Conjugation is one process by which atrazine and other pesticides are deactivated. In the process, the compound may become bound to a protein molecule where it is inactive.

Cultural Control

(Pest.) The use of crop tending methods other than pesticidal for minimizing harmful development of pests.

Cuticle

(Bot.) Waxy epithelium on surface of leaves.

DCPA

(Pest.) Dacthal; dimethyltetrachloroterephthalate. A pre-emergence herbicide used in nurseries, and elsewhere in agriculture.

Defoliant

(Pest.) A herbicidal compound whose principal effect is to remove leaves.

Defoliator

(Silv.) An insect that feeds on tree foliage.

Degradation

(Pest.) The process of destroying a pesticide residue. Most degradation is biological, resulting from metabolic action of plants and micro-organisms. Photolysis, chemical reduction and conjugation are also important degradation routes.

Desiccant

(Pest.) A compound used to dry plant material quickly. Dinoseb and paraquat are desiccants, as opposed to herbicides such as phenoxys or glyphosate that absorb, translocate and cause slower, systemic effects.

Dioxins, Dibenzodioxin

(Chem.) An organic compound having two carbon rings coupled by two oxygen bridges. There are many dioxins. Some of the dioxins made up of chlorinated benzene rings (the dibenzodioxins) are toxic contaminants of pesticides for which chlorinated phenols are used in manufacture.

Direct Effect

(Pest.) Effects of a pesticide occurring as the direct result of toxic action.

DNBP

(Pest.) Dinoseb; dinitro. Dinitro-o-secondary-butyl phenol. A desiccant used in forests exclusively to prepare for burning. This compound should be handled with care.

Dormant Spray

(Pest.) Application of a pesticide during the season when deciduous species have no leaves.

Drift

(Pest.) The movement of a pesticide with atmospheric flow, sometimes as vapor, but more often as fine droplet or dust.

Drift Control

(Pest.) Technology that reduces drift. Includes additives to increase viscosity or cohesion of spray mixtures, equipment modifications or other measures to reduce quantities of spray in five droplets, especially those smaller than 100μ

Duff

(Silv.) The organic layer of partly decomposed unrecognizable plant parts on the soil surface, between the litter layer and the A_1 horizon of soil.

Dust

Any finely pulverized, dry material. (Pest.) A pesticidal product prepared as a dust, to be applied by dispersal into an air stream.

Duster

(Pest.) A device capable of directing a high velocity air stream past a dust delivery system to form a cloud containing a known quantity of dust.

Ecosystem

(Biol.) The sum of all biological entities functioning in concert in a physical system.

Effective Swath Width (ESW)

(Pest.) The width of a swath as measured by the maximum distance between centers of contiguous swaths without leaving a strip of significantly lower deposit at the point of overlap. The ESW is much narrower than the total swath width of an aircraft.

Emergence

(Bot.) The emergence of germinating seed through the soil surface.

Emulsifier

(Pest., Phys.) A substance reducing surface tension of a liquid that is in contact with another so as to cause dispersal of one liquid in the other. Action is caused by an affinity of one end of the molecule for one liquid and the other end for the other, with the result that the emulsifier produces the largest possible surface interface between liquid phases, i.e. an emulsion.

Emulsifiable

(Phys.) The property of a liquid permitting dispersal in another liquid in minute droplets, forming a milky suspension of liquid particles in a continuous medium of carrier.

Emulsion

(Phys.) A mixture of two or more liquids in which at least two phases, insoluble in one another, are present, with one or more dispersed in droplets which are suspended in a continuous medium.

Equilibrium Reaction

(Chem.) A reversible reaction sequence in which a mixture of reacted and unreacted substances exist at the same time, and in which the proportion of reacted substance reflects the conditions in which the reaction is occurring. When the conditions change, the proportion of reacted substance changes in equilibrium with the conditions.

Ester

(Chem.) The reaction product of an acid with an alcohol. Among pesticides, phosphate esters are important as insecticides, and esters of phenoxys, dicamba, picloram and triclopyr are important bases for herbicide products. Esters are insoluble in water and soluble in oil.

Etiolation

(Bot.) The pale-colored slender-stalk elongation of plants in low light intensities.

Exposure

(Pest.) The actual contact of a metabolic system with a toxic substance. Physical external contact is not synonymous with exposure.

Extremely Toxic

(Pest.) Acute oral toxicity (< 5 mg/kg). See TOXICITY.

Flashback

(Silv. Pest.) The movement of an injected herbicide from a treated tree to an untreated tree so as to cause injury or death to both.

Flowable (Pesticide)

(Phys., Pest.) A liquid suspension of solid particles in which settling is inhibited by stabilizers. Used to formulate insoluble pesticides in liquid products.

Foliage Spray

(Pest.) Application of pesticide to foliage. Usually refers to treatment of woody species. Early foliage treatment is applied during leaf expansion; summer foliage spraying is done when most leaves are fully expanded and exporting carbohydrates. Late foliage season is when leaves are fully mature and terminal growth is complete. Fall foliage season is during the few weeks before leaf abscission begins.

Forb

(Bot.) A broadleafed herb.

Formulation

(Pest.) The preparation of a pesticidal substance in a form in which it can be used in the field. Formulation description specifies active ingredients, chemical form (i.e. ester, amine etc.), concentration, solubility. The formulation of a product often entails addition of surfactants, emulsifiers, solvents and other additives.

Fumigant

(Pest.) A gaseous pesticide.

Fungicide

(Pest.) A pesticide used to control fungi.

Gallery

(Silv.) The pattern in inner bark of a tree etched out by the feeding of larvae of bark beetles.

Granular

(Pest.) A pesticidal preparation in the form of sand-size particles of inert material coated with active ingredients. Normally applied dry. Is not retained by foliage, and is largely effective through root uptake.

Herbicide

(Pest.) A pesticide used to control plants.

Highly Toxic

(Pest.) Acute oral toxicity in the range of 5-50 mg/kg. (See TOXICITY)

High-Volatile

(Pest.) The term applied to ester herbicide formulations containing less than six carbons in the alcohol use for esterification. Ethyl, isopropyl and butyl esters are "high-volatile" esters.

Hill Reaction

(Bot.) The "light" reaction in photosynthesis in which the energy from photons of light is captured and converted into chemical energy, which is then used to reduce carbon dioxide into building blocks for carbohydrate synthesis.

Hormone-Type Herbicides

(Pest.) Herbicides that influence growth of plants so as to produce formative effects. Normally the result of effects on cell wall formation which cause curvature of elongating tissues when concentrations are uneven on two sides. Phenoxys, Picloram, triclopyr, dicamba and other benzoic acid herbicides are in this group. More properly described as synthetic growth regulators.

Hydraulic Sprayer

(Pest.) A sprayer of liquids. Usually applied to hose-gun sprayers used to apply large volumes of liquid to the point of runoff.

Hydroxy

(Chem.) An OH group attached to a molecule. (Pest.) Hydroxy atrazine or related compounds are the result of dechlorination and hydroxylation of the s-triazine compounds. The hydroxy-s-triazines are biologically inactive. The conversion is not reversible outside the laboratory.

Indirect Effect

(Pest.) Effects of pesticides attributable to change of habitat, and with negligible toxic involvement to the responding organism.

Injection

(Silv., Pest.) 1. Placement of chemical inside the bark of a tree with a cutting device equipped with a reservoir of chemical. 2. Placement of a chemical below soil surface with chisel or similar equipment rigged with spray nozzles on each plough shoe.

Insecticide

(Pest.) A pesticide used to control insects.

Intercalary Meristem

(Bot.) A group of cells immediately behind a plant growing tip in which all cell division occurs that leads to elongation.

Intermediate (Chemical)

(Chem.) A product of a chemical reaction that is used in another reaction. Term used in metabolic cycles consisting of series of reactions.

Invert Emulsions

(Pest.) An emulsion in which water is dispersed in a smaller volume of oil, with resulting creamy or mayonnaise consistency. These materials are used where high viscosity and large droplet size are desirable for reduction of drift.

IPM

Integrated pest management. (See PEST MANAGEMENT)

Label

(See PESTICIDE LABEL)

LC_{50}

(Pest., Toxicol.) The median lethal concentration of a toxic substance in food, for terrestrial animals or the medial lethal concentration in water for aquatic organisms.

LD_{50}

(Pest., Toxicol.) Median lethal dose of a toxicant administered to a population of test animals, expressed in milligrams toxicant administered per kilogram body weight. An acute LD_{50} is the median single oral dose, whereas a chronic LD_{50} is the median daily dose administered over an extended period.

Leaf Miner

(Silv.) An insect that feeds entirely within leaves.

Low Volatile

(Pest.) Describes esters of herbicides made with alcohols of six or more carbon atoms. Isooctyl, butoxy ethanol, propylene glycol butyl ether and butoxy ethoxy propyl esters are low-volatile products.

MAA

(Chem.) Methanearsonic acid. The acid of which MSMA is the monosodium salt and DSMA is the disodium salt.

METABOLITE

(Pest.) A compound formed by the breakdown of a pesticide by metabolic processes. May be more toxic than the parent compound, but is never more abundant than the original deposit and therefore seldom has greater toxic hazard. The metabolite of a pesticide is sometimes the compound that produces an effect in the target organisms. The ability to form the metabolite renders the target susceptible.

Microbial Insecticide

(Pest.) A preparation of bacterial or viral particles capable of transmitting disease to insects when dispersed as a spray.

Micron

A thousandth of a millimeter; a millionth of a meter.

Microsite

(Silv.) The immediate environment influencing a seedling.

Mist Blower

(Pest.) A machine for applying pesticides that directs an airblast past a nozzle enroute to a target so as to carry a fine mist or spray to the target. Produces maximum density of droplets per square inch on target, also maximum loss of fine droplet to evaporation, drift. See AIR BLAST MACHINE.

MMD

(Pest.) Mass median diameter. See VMD.

Moderately Toxic

(Pest.) Acute oral toxicity in the range of 50-500 mg/kg. (See TOXICITY).

Morphology

(Biol.) The external configuration of an organism; shape.

MSMA

(Chem., Pest.) Monosodiummethane arsonate. An herbicide used largely for tree injection. Also used in site preparation as an additive for phenoxy and other herbicides to enhance effects.

Mutagen

(Pest.) A substance that causes cellular genetic damage transferrable to offspring.

Nontoxic

(Pest.) Not having the ability to produce toxic effects at a level of dosage of 5,000 mg/kg. This level has been chosen arbitrarily, because virtually all substances can produce metabolic malfunction at some level of dosage. The probability of an organism absorbing enough of a material in this toxicity class to produce symptoms is remote.

Oncogen

(Toxicol., Pest.) A substance that causes formation of tumors in animals. May or may not be a carcinogen.

Organic Arsenical

(Chem., Pest.) Any compound having an arsenic-carbon linkage. Among pesticides, includes MSMA, DSMA, and Cacodylic acid (hydroxydimethylarsine oxide).

Organochlorine (Insecticides)

(Pest.) A group of insecticides characterized by bycyclic or polycyclic unsaturated structures and multiple chlorine substitution. Chemicals tend to be broad in spectrum and have considerable persistence. This group of compounds has an affinity for oil or fat, giving the various materials a tendency to accumulate with an extended period of exposure.

Organophosphorus, Organophosphate

(Chem., Pest.) Any organic compound containing a phosphorus atom or a phosphate group. An extremely large and useful group of insecticides, mostly with short residual life in the environment. Also includes a small number of herbicides.

Oxidized

(Chem.) The state of oxidation-reduction reactions having the lowest net energy content.

Parasite

(Pest.) An organism wholly or partly dependent on living hosts, but not contributing directly to the welfare of the host. A true parasite does not kill its host.

Parasitoid

(Biol.) An animal that feeds on or in another animal for a relatively long time, consuming most or all of its tissues, eventually killing its host.

Pellets

(Pest.) A pesticidal preparation usually consisting of a mixture of the pesticide with an inert organic filler that is compressed into solid particles of relatively uniform size, substantially larger than sand granules.

Persistence

(Pest.) The tendency of a compound to remain biologically active over a period of time after application. (See RESIDUE)

Pest

(Pest.) An organism or population of organisms that is too successful to be compatible with management objectives. A biological entity that is unwanted.

Pesticide

A substance of any kind used for controlling, damaging or repelling some species of pest. Includes insecticides, fungicides, herbicides, miticides, nematocides, etc. A generic term.

Pesticide Label

(Legal) A document or set of documents describing a registered pesticide. Includes all printed matter prepared by the manufacturer or its representatives, and in some instances consultation by representatives, regarding the product.

Pest Management

(Pest.) A system of resource management that minimizes interaction between pests and management systems through the integrated use of cultural, biological and chemical controls. Integrated pest management.

PGBE Ester

(Pest.) Propylene glycol butyl ether ester, usually of a phenoxy herbicide. One of the low volatile esters.

Phenoxy Herbicides

(Pest.) Any herbicidal product with active ingredient chemistry based on a chlorinated phenyl group connected to an aliphatic acid by an oxygen link.

Pheromone

(Biol.) A sex-related attractant for insects that acts through olfactory senses.

Phosphonoalkyl

(Chem., Pest.) A group of compounds in which an oxidized phosphorus group is linked to an organic compound by a P-O-C linkage. Includes a group of herbicides of which an important representative is glyphosate.

Pioneer

(Bot.) A plant that colonizes recently disturbed sites.

Post-Emergence

(Pest.) Refers to timing of pesticide application during the several weeks immediately following emergence.

Predator

(Biol.) An animal that kills animals for food. Generally many individuals are consumed during the life of the predator.

Pre-Emergence

(Pest.) The timing of pesticide application before seedling emergence occurs.

Prescription

(Pest., Silv.) The sum of all components toward a remedy for a set of biological problems.

Primary Effect

(Pest.) In a toxic substance, the metabolic effect observed at the lowest "effect" dosages in target species. Usually masks other effects, if any, because lethality may occur at dosages below those affecting other metabolic systems.

Radical

(Bot.) The root tip of a germinant.

Reduced

(Chem.) The state of oxidation-reduction reactions having the highest energy content.

Release

(Silv.) Selective control of undesirable vegetation in the presence of desirable species.

Residual Hardwoods

(Silv.) Standing hardwoods remaining after logging. Usually of low quality.

Residual Pesticide

A pesticidal substance that retains some ability to control the target pests for an extended period after application.

Residue

(Pest.) The remaining pesticide at some given time interval since application. Residue is determined by rate of application and rate of degradation. Residue also refers to pesticide remaining in food crops, which may be compared to allowable tolerances to determine whether food can be sold in interstate commerce.

Roller Pump

(Pest.) A rotary pump in which the vanes are rollers, and impeller rotation is in an eccentric chamber.

Salt

(Pest., Chem.) An ionic compound consisting of an anion and cation. The water-soluble formulations and phenoxys, picloram, dicamba, triclopyr and glyphosates are mostly amine salts. Any reaction product of an acid with a base.

Sanitation Salvage

(Silv.) The programmed harvesting of commercial timber species in which trees with high risk of insects and diseases are given priority for removal. System is used principally in overmature forests in the Western Pine Region.

Sawfly

(Silv.) A defoliating forest insect that causes damage to conifers. Order Hymenoptera.

Scalping

(Silv.) The practice of removing, by physical means, the surface soil and vegetation in the vicinity of a planted seedling.

Secondary Effect

(Pest.) In a toxic substance a metabolic reaction that may occur at elevated dosage levels in organisms not subject to primary effects.

Seed Orchard

(Silv.) A plantation of trees established solely for production of seeds.

Seral

(Bot.) Early in succession after some disturbance.

Shade Leaves

(Bot.) Leaves formed in shady locations with structural adaptation for photosynthetic efficiency in low light intensity. Normally injured severely by sudden exposure to full sunlight.

Shothole Feeder

(Silv.) An insect that feeds on leaves, leaving more or less round holes in leaves.

Site Preparation

(Silv.) The preparation of forest land so as to create suitable conditions for regeneration of desirable species, usually in plantations.

Skeletonizer

(Silv.) An insect that feeds on foliage and leaves the network of vein structure and usually one surface of the leaf.

Slightly Toxic, Low Toxicity

(Pest.) Acute oral toxicity in the range of 500-5,000 mg/kg. (See TOXICITY)

Solubility

(Phys.) The ability for a substance to mix with a solvent (such as water, oil or other liquid), so as to form a clear solution.

Solubilized Acid

(Pest.) A pesticide compound of low solubility rendered soluble or emulsifiable by mixing in an organic solvent combined with an emulsifier and surfactants.

Soluble

(Chem.) A property of a substance indicating that it may be dissolved in a liquid to form a clear, transparent and homogeneous mixture that will not settle out.

Stagnation

(Silv.) The reduction in growth of individual trees when stand density is excessive.

Stand Improvement

(See TIMBER STAND IMPROVEMENT

S-Triazine

(See TRIAZINE)

Succession

(Biol.) The sequence in which plants and other organisms replace one another after disturbance, leading eventually to a climax community in the absence of further disruption.

Sucking Insect

(Silv.) Insects that feed on leaf or phloem tissue by insertion of proboscis through bark or leaf surface.

Sun Leaves

(Bot.) Leaves formed in full sunlight. Characterized by deep spongy inesophyll and thick epidermis, but low photosynthetic capacity when placed in shade.

Sunscald

(Silv.) The necrosis in bark or foliage caused by heat from direct sunshine. Usually caused by opening of stands so that unprotected stems with thin bark and dark color are directly exposed to sun at a low angle, and where protected from cooling effects of wind. Occurs after thinning or hand release of conifers.

Surfactant

(Pest.) A substance added to a spray mixture to cause it to adhere to a surface by decreasing surface tension.

Swath

(Pest.) The area covered by a single pass of a pesticide application device. See EFFECTIVE SWATH WIDTH.

Symbiont, Symbiosis

(Biol.) An organism that depends wholly or in part on a host organism and which provides some benefit to its host. The process of interchange of benefits between host and symbiont.

Systemic

(Pest.) A pesticide that translocates readily in a plant or animal wherein such translocation enhances pesticide utility.

Target Species

(Pest.) A species for which control is desired in a pest management program. All other species are non-target species.

TCDD

(Chem.) One or more of 22 isomers of tetrachlorodibenzodioxin. The 2,3,7,8 isomer occurs as an unavoidable contaminant of 2,4,5-trichlorophenol in trace amounts. It is formed when tetrachlorobenzene is reacted under alkaline and high pressure conditions to form trichlorophenol. Quality control in manufacture has all but eliminated TCDD in Silvex, 2,4,5-T and other pesticides. TCDD is extremely toxic in pure form. It is also a by-product of combustion.

Threshold

(Toxicol.) An entry point. In the toxiocological context, the entry into the range of doses producing observable effects, i.e. the lowest "effect" dosage.

Timber Stand Improvement

(Silv.) The set of operations in an existing stand directed toward an improvement in spacing, quality or species composition in accordance with management objectives. Usually in stands between stages of established regeneration and commercial thinning.

Tolerance

(Pest) Resistance to pesticide

(Silv.) The ability of a plant to persist in understory conditions. Traditionally defined as ability to withstand shade, tolerance appears to include ability to withstand mechanical injury from browsing or falling debris.

Tipkill

(Silv., Pest.) The partial kill of a plant by an herbicide. Usually refers to woody plants in which the roots and perhaps parts of stems remain viable sources of sprout regeneration.

Toxic

(Pest.) Having the ability to produce damage to metabolic systems as the result of a specific exposure level. (See TOXICITY)

Toxic Hazard

The likelihood of occurrence of toxic effects. Toxic hazard exists *only* as the result of significant *exposure* to a substance of significant *toxicity*. A great many substances are toxic, but are not physiologically encountered in toxic quantities and therefore do not pose toxic hazard.

Toxicity

(Toxicol.) The ability for a substance to produce deleterious metabolic effects in an organism. Degree expressed in terms of milligrams dosage per kilogram body weight. Median lethal dose (LD_{50}) is a numerical measure of the dosage needed to cause mortality of 50 percent of a test animal population. Degrees of toxicity are:

Extremely toxic – LD_{50} < 5 mg/kg
Highly toxic – LD_{50} 5-50 mg/kg
Moderately toxic – LD_{50} 50-500 mg/kg
Slightly toxic – LD_{50} 500-5,000 mg/kg
Non-toxic – LD_{50} > 5,000 mg/kg

Trade Name

Proprietary name of a product.

Translocate

(Pest.) Refers to movement of a pesticide in plants, animals or soils to some site of metabolic action.

Triazine

(Chem.) An organic compound with three nitrogen atoms in a 6-membered ring structure. Triazine pesticides include simazine, atrazine, hexazinone, and many others.

TSI

(See TIMBER STAND IMPROVEMENT)

Ultra-Low Volume

(Pest.) Refers to dispersal of pesticides with very little carrier; total delivery per unit area is usually below a half-gallon per acre, (5 l/ha).

Vane Pump

(Pest.) A rotary pump in which the impeller includes flexible vanes, moving around in an eccentric chamber during operation. Flexing of the vanes, and changing volume between vanes, results in displacement of liquid. Normally a moderately high volume, self-priming, moderate pressure device.

Vigor

(Silv.) Describes the ability of a tree to grow. Trees of high vigor have full, deep crowns and can achieve maximum growth rate. Trees of low vigor usually have thin, short or off-color crowns, and are likely to display poor growth and susceptibility to various pests and physical impacts. Vigor is related to the ratio of foliage to total biomass.

VMD

(Pest.) Refers to the droplet size in a spray. Specifically, the diameter of a droplet with median volume

Volatile

(Phys.) Having the ability to evaporate.

Vortex

(Pest.) The pattern of air turbulence following an aircraft. The vortex has an important effect on distribution of aerially applied pesticides.

Weed

(Pest.)A plant growing where it is not wanted. Includes herbs, shrubs and trees, and may include specimens of desirable species, e.g. where stands are overstocked, superfluous stems are weeds.

Wettable Powder

(Pest.) A finely pulverized dust prepared for mixing with water. Usually of low water solubility.

Whole Leaf Feeder

(Silv.) A leaf-eating insect that consumes the entire leaf.

INDEX

Absorption, 24, 26, 27, def. 190
Active Ingredients, 44, 46, 48, 50, def. 190
Acute Toxicity, def. 191
Adjuvant, def. 191
Adsorption, 24, def. 191
Aerial Application, 60, 62, 73
Agitation, 131, def. 191
Air Blast Machine, 138, def. 191
Aircraft, 139-142
  description, 139-140
  guidance of, 140
  swath, 141
Alder, 12, 13, 14, 16, 61
Amine, 29, 30, def. 191
Amino Acid, def. 191
Amitrol, 22, 39 (illus.), 44, 45, 51, 53, 58, 130
AMS, 22
Animal Damage, 13, 16, 55, 69-70
Ants, Carpenter, 110
  Texas leaf-cutting, 118
Aphid, 148
  balsam twig, 91
  woolly elm, 92
  poplar petiole gall, 92
Application, 27, 31, 44, 46, 48
  method, 50-53, 62-68
  aerial, 73
  season of, 56, 57, 60, 61
  technology, 62, 127-146
  equipment, 138-146
Appropriate Silviculture, 5, 18
Arsenicals
  inorganic, 65, 68, 114
  organic, 36, 37, 65, 67-69, 71, 72, 146; def. 191
    properties, 155-156
  metabolism, 155-156
  poisoning treatment, 169
Ash, White, 60
Asulam, 22, 39 (illus.), 44, 45, 51, 53, 58
Asymmetric Growth, 31, def. 191
Atrazine, 22, 33, 34 (illus.), 40, 44, 45, 51, 53, 56, 57, 58, 64
Attractants, 98
Bacillus Thuringiensis, 120, 121
Baits, 127
Bark Beetles, 81, 83-84, 104-107, def. 192
  control, 37, 123
  table of, 106, 107
Bark Sprays, 29, 32
Basal Spray, 64, 65, def. 192
Be Ester, def. 192
Beetle
  Ambrosia, 88, 110, 113
  Caragana blister, 92
  Douglas-fir, 84
  Engraver, 84, 88
  European elm, 104
  Longhorned, 110
  Mountain pine, 85, 105
  Powder post, 88, 110
  Southern pine, 81, 104
  Spruce, 84
  Western pine, 83
Bifenox, 22
Bioaccumulation, 149, 152
Biocide, def. 192
Biomagnification, 149, 152
Bipyridinium ion, def. 192
Birches, 13, 60
Birds
  effect of management on, 69
  effect of pesticides on, 149
Black Locust, 12
Blue Stain, 88
Boom, def. 192
Borers
  Boxelder, 92
  Flatheaded, 88
  Locust, 93
  Marine, 89
  Poplar, 110
  Round headed, 88, 110
  Sugar maple, 90
Bracken, 39, 50, 51
Broadleaf Weed, def. 192
Browse, 16, 36, 68
Brush (*see* target spp.), 50-53
Budworm
  Spruce, 81, 85, 91, 94, 117, 122
  Western spruce, 122
Buffers, Streamside, 166-167
Butterfly, Pine, 80, 85
Cacodylic Acid, 22, 37, 66
Calibration, 27
Cankerworm, Fall, 92
Canopy, def. 192
Carbamates, 117-118, 152, def. 192
  poisoning, treatment of, 171
Carbaryl, 117, 118, 152
Carcinogen, def. 192
Carpenter Worm, 108, 112
Carrier, 44, 46, 48, 57, 59, def. 192
Caterpillar, Tent, 81, 90

Ceanothus, 12
Certified Applicator, def. 192, 193
Chemical Thinning, 60, 88, def. 193
Chlorinated Hydrocarbon, 115, 116, 148, 149, def. 193. *See also* Organochlorine.
  poisoning treatment, 172-174
  properties, 148
Chloropicrin, 29, 42
Cholinesterase Inhibition, 149, 152, def. 193
Christmas Trees, 50, 51, 79, 90-91, 148
Chronic Toxicity, 149-151, def. 193
Cinerins, 115
Cleaning, Silvicultural, 56, def. 193. *See also* Release; hand release.
Clothing, Applicator, 72, 139, 153, 159-160
Coleoptera, 78
Common Name, 23, def. 193
Cone and Seed Insects, 101-104
  table 102-103
Coneworms, 90
  Southern pine, 81
Conjugation, def. 193
Costs, 45, 47, 49
Coverage, pesticide, 128, 130-135
Cryolite, 114
Cultural Control, 81, 82, 83, 85, 86, def. 193
Cuticle, def. 194
Dalapon, 22, 39 (illus.), 40, 44, 45, 51, 58, 64, 130
Damage
  animal, 70
  conifer, 63
  crop, 72
  insect, 79
DCPA, 22, 41, def. 194
DDT, 115
Debarker, 88, 110
Debris, 59. *See also* Litter.
Defoliant, def. 194
Defoliator, 84-86, 94-98, def. 194
Degradation, 149, def. 194. *See also* persistence; microbial degradation.
Desicant, 32, def. 194
Desiccation, pre-burn, 32, 63
Dicamba, 22, 37, 38 (illus.), 44, 45, 51, 53, 56, 58
Dichlorprop, 22, 30 (illus.), 31, 54
Difluorobenzuron, 120
Dinitrophenols, 174
Dinoseb, 22, 40 (illus.), 44, 45, 51, 53, 154
  properties, 154, 155, 159, 160
  poisoning, treatment of, 174-175
Dioxin, Dibenzodioxin, 33, def. 194. *See also* TCDD.
Diphenamid, 22, 41
Direct Effect, def. 194
Disturbance, 73
Diversity, Species, 68, 69
DNBP, def. 194
Dominance; Domination, 11, 15, 56, 74
Dormant Spray, 32, 63, 64, 69, def. 194
Douglas-fir, 14, 15, 16, 59, 61, 88
Drift, 29, 62, 64, 72, 73, 134-137, 161, 162, def. 194
  control of 134-137
Drop Size, 62, 63, 135-136, 161
Drought, 16, 55, 84. *See* Moisture stress.
Duff, def. 195
Dusts, 139, def. 195
Duster, 139, def. 195
Dutch Elm Disease, 104
Economic Injury, 79
  threshold, 80
Ecosystem, def. 195
Education, 3
Effect, Indirect, def. 198
Effective Swath Width (ESW), def. 195
Emergence, def. 195
Emulsifiable Concentrates, 28, def. 195
Emulsifiers, 29, 128, def. 195
Emulsion, 28, 128, 130, 131, def. 195
  invert, 130
Endosulfan, 148-149, 160
Endrin, 148-149
Equilibrium Population, 80
Equilibrium Reaction, def. 196
Equipment, Spray, 60, 62, 138-146
Ester, 29, def. 196
Etiolation, def. 196
Exposure, human, 67, 71, 89, 144, 146, 147, 149, 156, 157, 158-161; def. 196
  chronic, 149
  dermal, 72, 158-159
  herbicidal, 157
  phenoxy herbicide, 159
Extremely Toxic, def. 196 *See* Toxicity
Farming, Organic, 162
Ferns, 50, 51
Filters, 132
Fir, true, 58, 59, 85
Fire, 20, 21, 55, 155
Fish
  herbicide effects, 156-157

insecticide effects, 151, 152, 153
kills, 164
Flashback, 60, 67, def. 196
Flies, 92
Flowable (pesticide) 25, 27, 30, 130, def. 196
Foliage Spray, 60-63, def. 196
Forb, def. 196
control of, 50, 51
Formulation, 24-30, 44, 46, 48, 60, def. 196-197
Fosamine, 22, 35, 36, 46, 47, 51, 53, 58, 63
Fumigant, 25, 29, 42, 114, 118, def. 197
poisoning treatment, 177-178
Fungicide, def. 197
Gallery, def. 197
Game, Big, 69, 70
Girdling, 74
Glossary, 190-207
Glyphosate, 22, 35 (illus.), 36, 46, 47, 51, 53, 54, 56, 57, 58, 63, 135
Gophers, 12
Granular, 25, 27, 30, def. 197
Grass,
control, 11, 50, 51, 58
habits, 11
Grasshoppers, Spurthroated, 92
Growth Regulator, 119-120
Habitat, 15, 18
effect of change, 154
management of, 6, 9, 68-70
Hand Release, 56, 59
hazards of, 71
Hardwoods, 52, 53, 54, 56, 58, 60, 65
cull management, 65-67; 69, 73
sensitivity to herbicides, 54, 56, 58
Hazards
humans, 157-161
insect, 84
non-target species, 158, 161-167
wildlife, 71
Hazard, Toxic, 65, 71
Helicopters, 140-142
Hemlock, 58, 59, 61, 88
Herbs, 11, 16, 20, 21, 50, 51, 55, 58
Herbicide
properties, 30
silvicultural use, tables 22, 23, 25, 44-53
toxic hazard, 153-157
Hexazinone, 23, 33, 34 (illus.), 46, 47, 51, 56, 58, 64, 68, 130
Hickories, 13, 14
Highly Toxic, def. 197. *See* toxicity.
High-Volatile, def. 197
Hill Reaction, def. 197
Hormone-Type Herbicides, def. 197
Horntails, 110
Hydraulic Sprayer, def. 197
Hydrocarbons
chlorinated, 115-116, 148, 149
poisoning treatment, 172-174
non-chlorinated
poisoning treatment, 175-177
Hygiene, personal, 153, 160, 161
Impacts, *See also* Non-target species, 17, 18, 19
Indirect Effect, def. 198
Injection, 26, 30, 60, 65, 66, 67, 144-146, def. 198
Injury, Conifer, 57, 60, 61
economic, 79
Insects and their control, 77-126
Insecticides, 2, 112-135, def. 198
application, 114, 122
nomenclature, 119
organic, 115-125
table of sensitive pests, 124
properties, table 150-151
systemic, 114
Intercalary Meristem, 136, def. 198
Intermediate (Chemical), def. 198
Invert Emulsion, 130, def. 198
IPM, def. 198. *See* Pest management.
Keen's Classification, 83
Label, 22, 23, def. 198
Lacrimators, 177
Laws, 19
LC_{50}, 164, def. 198
LD_{50}, 150, 154, 164, def. 198-199
Leaf-eating Insects, 94-98
table, 96
Leaf Miner, 95, def. 199
Liberation, 56. *See also* Release.
Lindane, 116, 149
Litter, Effects of, 16
Loblolly Pine, 14, 15
Locust, Black, 12, 93
Lodgepole Pine, 15, 84
Logging, 9, 13, 7
Low Volatile, def. 199. *See also* Volatility.
MAA, def. 199
Machinery (Heavy), 19, 20, 21, 55
hazards of, 71
Malathion, 117
Management Objectives, 3, 18, 43
Management Plans, 74, 75, 76
Manual Brush Control, 20, 21, 57, 59, 71
Manzanitas, 13
Maples, 13, 14, 17, 60
sugar, 90

Mercurials, 114
Metabolite, def. 199
Methoxychlor, 116
Methyl bromide, 29, 42, 118
Microbial Degradation of Pesticides, 149. *See* degradation.
Microbial Insecticide, 120, 126, def. 199
Microfoil, 141
Micron, def. 199
Microsite, def. 199
Midges, 148
Mineral Spirits, 23, 41
Mine Spoils, Reclamation of, 92-93
  insects of, 92-93
Mist Blower, 62, 138-139, 143, def. 199
Mites, 121
Mixing of Pesticides, 127-130
  wettable powders, 131
MMD, def. 199. *See* VMD.
Moderate Toxicity, 199. *See* toxicity.
Moisture Stress, 16, 55, 84
Molting Disruptants, 120
Monterey Pine, 15
Morphology, def. 199
Mortality, Seedling, 15
  tree, 79
Moth
  cone, 81
  Douglas-fir tussock, 94, 98, 120
  Gypsy, 81, 120, 122
  Nantucket pine tip, 91
  Pandora, 80
MSMA, 23, 37, 46, 47, 54, 65, 66, 88, def. 200
Mutagen, def. 200
Napropamide, 23, 41
Nicotine, 115
Nitrogen Fixation Management, 12
Noble Fir, 15
Non-chemical Weed Control Methods, 11
Non-target Species, Pesticide Effects, 161-167
Nontoxic, def. 200
Northern Weed spp. Susceptibility, 54
Nozzle Systems, 62
  description, 132-134
  function, 130-131
Nurse Crops, 12
Nursery, 41
Oakworm
  organstriped, 80
  redhumped, 92
Oil (Carrier), Use of, 28, 57, 59, 64, 65, 128
Oil-soluble Liquids, 28, 29
Oncogen, def. 200
Organic Arsenical, 65, 67, def. 200
  poisoning, treatment of, 169-170
Organochlorine (Insecticide), def. 200. *See* chlorinated hydrocarbon.
  accumulation of, 149
Organophosphorus, Organophosphate, 116-117, 152, def. 200
  treatment of poisoning, 178-180
Paint, Protective, 89, 110
Paradichlorobenzene, 118
Paraquat, 23, 40 (illus.), 41, 156
  toxicology of, 156, 159, 160
Parasite, 79, 121, def. 200
Parasitoid, 79, def. 201
Parathion, 116-117
Pellets (Formulation), 26, 28, 30, 67, 68, def. 201
Persistence, 45, 47, 49, 149, 150, 152, def. 201
  insecticides, 150
  herbicides, 154
Pest, 2, 3, 5, def. 201
Pesticide, 2, 3, def. 201
Pesticide Label, 22-23, def. 201
Pest Management, 6, 81, def. 201
PGBE Ester, def. 201
Phenology, 61
Phenoxy Herbicide, 29, 30, 57, 63, 64, 69, 135, 159, 161, def. 201
  effects on humans, 159
  toxicology of, 156-157
  treatment of poisoning, 180-182
Pheromone, 98, 121, def. 201
Phloem Insects, (Table) 108-109
Phosphamidon, 117
Phosphonoalkyl Herbicides, 35, def. 202
Photodecomposition, 115
Physiology, Human, 153
Picloram, 23, 29, 37, 38 (illus.), 46, 47, 51, 53, 54, 56, 58, 63, 64, 65, 66, 67, 72, 135
Pine
  loblolly, 14, 15
  lodgepole, 15, 84
  ponderosa, 12, 13, 15, 32, 61, 84
  slash, 15
  southern yellow, 14, 15, 16, 32
  sugar, 58
  white, 15, 16, 87
Pioneer, def. 202
  species, habits of, 10-14
Plans, Management, 73-75, 86
Plantation Management, 10-17, 86
Poisoning, Human, 152, 153, 158-160,

169-190
management of, 169-190
treatment of, 160, 169-190
Ponderosa Pine, 12, 13, 15, 32, 61, 84
Poplars, 13
Post-emergence, 41, def. 202
Predator, 79, def. 202
Pre-emergence, 41, def. 202
Prescription, 4, 42, 43, 70, def. 202
Primary Effect, 148, def. 202
Products Insects, 87-89, 110-113, table 111
Pronamide, 23, 39 (illus.), 40, 48, 49, 64
Pumps, Description, 132, 203, 207
Pyrethrins, 115
poisoning treatment, 182-183
Pyrethroids, 115
poisoning, treatment, 182-183
Pyrethrum, 115
poisoning, treatment of, 182-183
Rainfall, Effect On Chemicals, 64. *See* drought.
Recreation Areas, Insects, 92
Redwood, 15
Reforestation, *See* Regeneration.
Regeneration (of trees), 11, 13, 16, 17, 75
Registration, 22. *See* Label.
Release, 17, 50-53, 56-60, 64, 75, def. 202
Residual Hardwood, def. 202
Residual Pesticide, 24, 64, def. 202
Residue, Pesticide, def. 202-203. *See also* Persistence.
in water and soil, 154, 162
Resistance (To Herbicides), 53, 55
Rights Of Way, 17
Root Insects, 100-101
table, 100-101
Rotenone, 115
poisoning, treatment of, 183-185
Saddled Prominent, 90
Safety, 33, 70-73, 147-170
factors, 148, 149
Salmonberry, 16, 61
Salt, def. 203
Sanitation Salvage, 83, 105, def. 203
Sap-sucking Insect, 87
Sawfly, 91
European pine, 91
Virginia pine, 95
Yellowheaded spruce, 91
Scalping, 55, def. 203
Scots Pine, 13
Seasonal Effects of Herbicides, 32, 41, 42, 44, 46, 48, 52, 53, 56, 57, 61, 63
Secondary Effect, 148, def. 203
Seed and Cone Insects, 101-104, table 102-103
Seedbed, 73
Seedlings, Characteristics of, 13, 74
Seed Orchard, def. 203
insects of, 89-90
Selectivity
herbicide, 45, 47, 49, 61
insecticide, 120, 165
Sensitive areas, 75
Shade Leaves, 59, def. 203
Shade Trees, 79
insects of, 89
Shading, 13, 15, 16, 57, 59
Shoot Insects, Table 100-101
Shothole Feeders, 95, def. 203
Shrubs, 50-53
Silvex 23, 30 (illus.), 31, 48, 49, 51, 53, 58, 64, 65, 157
Simazine, 23, 33, 34 (illus.), 48, 49, 51, 57, 58, 64
Site Preparation, 17 (def.), 50-53, 55, 75, def. 203-204.
Skeletonizer, 95, def. 204
Skin, Exposure To Pesticides, 72, 158-159
Slash Pine, 15
Slightly Toxic, Low Toxicity, def. 204. *See* toxicity
Snowberry, 13
Sodium Arsenite, 65, 68
Soil-applied Herbicides, 57
Soil Disturbance, 11
Soil, Uptake From, 30
Soil, Pesticide Persistence. *See* Persistence.
Solubility of Pesticides, 25, 57, 148-149, def. 204
Solubilized Acid, def. 204
Southern Weed spp, Susceptibility, 56
Southern Yellow Pines, 14, 15, 16, 32
Spanworm, Elm, 85
Spills, Cleanup, 71-72
Spittle Bug, Saratoga, 87
Spot Treatment, 71
Sprayers, 62, 131-134, 138-139, 142-144
Sprouting, 13, 70
Spruces, 15, 32, 58, 59, 85
Stagnation, def. 204
Stand Improvement,
See Timber stand improvement

Streamside, Buffers 166-167
habitat, 166-167
Stump, Treatment of, 67, 68, 146
S-Triazine,
See Triazine
Succession, 16, def. 204
Sucking Insects, 98, def. 204
Table, 99
Sugar Pine, 58
Sulfur Compounds, 114
lime, 114
poisoning, treatment of, 185-186
Sun Leaves, def. 204
Sunscald, 59, 67, def. 204-205
Suppression, 13, 16, 56, 57
Surface Tension, 128
Surfactant, 26, 135, def. 205
Suspensions, 131
Swath, Descr., 141, def. 195, 205
width, 141, 166, def. 195
Symbiont, Symbiosis, def. 205
Systemic, 26, 31, 36, def. 205
Target Species, 45, 47, 49, 50-53, def. 205
TCDD, 33, 157, def. 205
Termites, 89, 110
Thinning, Pre-commercial, 67
Threshold, def. 205
economic, 80
toxicity, 158
THUJA Spp, 15
Timber Stand Improvement, 6, 60, def. 205
Tolerance, 13, 15, 16, def. 205-206
Topkill, def. 206
Toxaphene, 160
Toxic Hazard, 65, 71, 152, def. 206
Toxicity
acute
herbicides, 153-157
insecticides, 150-151, 152
chronic, 149, 151
human, 151, 155, 158-160
non-target, spp. 164
standards, degree, 206
Toxicology, 147-148
Trade Name, 22, 23, def. 206
Translocation, 26, 31, 36, def. 206
Triazine, 33, def. 206
poisoning, treatment of 186-187
Triclopyr, 23, 38 (illus.), 48, 49, 51, 53, 54, 58, 65, 66, 69
TSI, def. 206. *See* Timber Stand Improvement.
2,4-D, 23, 30 (illus.), 31, 33, 48, 49, 51, 53, 54, 56, 58, 64, 65, 68, 69, 72, 159
poisoning, treatment of, 180-182
2,4,5-T, 23, 30, (illus.), 31, 33, 48, 49, 51, 53, 54, 56, 58, 63, 65, 69, 157, 159
poisoning, treatment of, 180-182
Ultra-Low Volume, def. 206-207
Uptake, Route of, 45, 47, 49
Urban Pest Control, 89
Utilization, Weed, 17, 18
Vane Pump, def. 207
Vegetation Management, 9-76
Vigor, Tree, 82, 83
Virus, Nucleopolyhedrosis, 120-121
Viscosity, 128, 130
VMD, def. 207
Volatility, 29, 162, def. 207
high volatility def. 197
low-volatility def. 199
Volume, Carrier, Effect of, 135
Vortex, 141, def. 207
Walking Stick, 98
Water, quality, 75, 149, 152, 153, 162
criteria, 165
management, 163-167
monitoring, 167
tables of factors concerning, 163, 164-167
Weather, Effects of, 85
Weed, 5, 9-75, def. 207
Weeds, characteristics of, 10, 11-13
Weed Control Methods, 18
Weed Trees, 52, 53
Weevil, Terminal Shoot, 81, 87, 91, 100, 103
Western Weed Spp. Susceptibility
conifer, 54, 56, 58, 61
hardwood, 54, 56, 58, 61
Wettable Powder, 25, 27, 29, 30, 130, 131, def. 207
mixing, 131
White Pines, 15, 16, 87
Whole Leaf Feeder, 95, def. 207
Wildlife,
effects of habitat changes 68-71, 154
hazards of pesticides, 71, 157-158, 161-167
Windbreaks, Insects of, 91-92